Agricultural Science: Gene Sequencing and Mapping

Agricultural Science: Gene Sequencing and Mapping

Shashi Shekhar Prasad

RANDOM

RANDOM PUBLICATIONS

NEW DELHI (INDIA)

Agricultural Science: Gene Sequencing and Mapping

ISBN 978-93-5111-585-4

Published in 2015 in India by

RANDOM PUBLICATIONS

4376-A/4B, Gali Murari Lal, Ansari Road
New Delhi-110 002
Phone : +9111-43580356, 011-23289044, 011-43142548
e-mail: sales@randompublications.com,
info@randompublications.com, randomexports@gmail.com

Reprinted 2018

Type Setting by : Friends Media, Delhi-110089
Digitally Printed at : Replika Press Pvt. Ltd.

Preface

Agricultural science is a broad multidisciplinary field of biology that encompasses the parts of exact, natural, economic and social sciences that are used in the practice and understanding of agriculture. With the exception of theoretical agronomy, research in agronomy, more than in any other field, is strongly related to local areas. It can be considered a science of ecoregions, because it is closely linked to soil properties and climate, which are never exactly the same from one place to another. Many people think an agricultural production system relying on local weather, soil characteristics, and specific crops has to be studied locally. Others feel a need to know and understand production systems in as many areas as possible, and the human dimension of interaction with nature. Agriculture sciences seek to feed the world's population while preventing biosafety problems that may affect human health and the environment. This requires promoting good management of natural resources and respect for the environment, and increasingly concern for the psychological wellbeing of all concerned in the food production and consumption system. Economic, environmental, and social aspects of agriculture sciences are subjects of ongoing debate. Recent crises illustrate the complexity and importance of this debate.

I would like to thank my team for standing beside me throughout my career and writing this book. My special thanks go to "Random Publications" who have published the book.

– Shashi Shekhar Prasad

Contents

1

Biotechnology Developments in Agricultural Sector

The Department of Biotechnology (DBT) established under the Ministry of Science and Technology in 1986 was the major instrument of action to bring together most talents, material resources, and budgetary provisions. It began sponsoring research in molecular biology, agricultural and medical sciences, plant and animal tissue culture, biofertilizers and biopesticides, environment, human genetics, microbial technology, and bioprocess engineering, etc. The establishment of a number of world class bioscience research institutes and provision of large research grants to some existing universities helped in developing specialized centres of biotechnology.

There exists a gradient of biotechnologies, depending on the degree of sophistication, complexity, stage of development and application. The lower side of the gradient comprises simpler but widely used techniques such as in vitro culture, rhizobium technology, fermentation. The upper side of the gradient includes advanced techniques involving genetic engineering. The gradient of biotechnologies may be matched with the gradient of national capabilities, economic investment and efforts to provide the possibility of choosing the appropriate techniques and approaches with the most positive impacts. The pattern of use and development of biotechnologies in developed and developing countries thus varies considerably.

BIOTECHNOLOGY IN AGRICULTURAL DEVELOPMENT

We are at the threshold of 21st century, where great challenges on three fronts namely, production, pollution and population are awaiting the developing countries. Moreover, in the era of privatization, globalization and liberalization, there is heavy international competition within as well as between the countries on all fronts including the agriculture. The present scenario indicates that there would be "survival of the fittest". Needless to mention that by 2000 AD our population will be around one billion the priority is clear-cut before us. We have to climb upward to achieve higher and higher production to ensure that every mouth of our ever increasing population has enough to eat. The quest for agriculturists for greater productivity and improvement of existing cultivars with better quality food continues with great vigour and spirit. The conventional methods of plant breeding and traditional agricultural practices have done a tremendous job and contributed to a great deal towards the above goal.

However, in view of the acuteness of the problem and renewed fears regarding the availability of the proper and enough food, these methods alone are not sufficient to meet the situation.

Although, the FAO believes that food production will continue to increase for the next 20 to 25 years at the same rate as experienced during the past 30 years to meet the futuristic demand. The worldwatch Institute has, however, reported that this rate may not be sustainable.

Moreover, they believe that science and technology can no longer ensure the onward march of achieving higher and higher yields. In addition we are also subjected to an almost daily litany of doom and despair as a result of global warming, depletion of ozone layer, oil and water running out and so on.

The next century certainly packed with trouble if the meteor does not get us first, then we will gradually either be roasted, frozen or desiccated, and to make doubly sure, shall certainly be starved as well. It seems that if we have to stand any chance whatsoever in the next century, our capacity to adopt the living resources will be extremely important for our survival.

Because of changes in the environment and depleting energy resources, this capacity to adapt will be every bit as important to those areas of the world which are blessed with overproduction like Europe today. Accelerated adaptation through genetic change is, of course, one of the important means by which plant breeders have achieved their aims in the past.

Today the new techniques of biotechnology in general and genetic engineering in particular may offer to the plant breeders and agriculturists the chance of speeding up adoption to an extent hitherto considered impossible. This new power will undoubtedly bring risks but it could also be the only chance we have for escaping the dangers ahead.

Biotechnology being a multidisciplinary subject requires coordinated efforts and expertise of scientists with different backgrounds to put together in order to achieve success. It may be defined as may technique that uses living organisms, or parts of organisms to make or modify products, to improve plants or animals or to develop microorganism for specific uses.

It includes well established technologies such as those used in breeding, plant propagation and conventional animal vaccine productions. Modern biotechnology encompasses the more recently developed techniques involving the use of recombinant DNA technology, monoclonal antibodies and new techniques of cell culture. Recent development in biotechnology have made it possible to move genes from microbes, plants and animal species into plants of interest. Biotechnology helps in the commercial utilization of gene and other things for the benefit of human kind. However, destruction of ecosystems has been a major concern of biotechnology. Developing countries are naturally attracted to the potential applications of biotechnological research in solving problems of hunger, energy supply and improving the quality of life.

The priorities of different countries however very widely, In this context, the National Biotechnology Board of India has chosen genetic engineering, photosynthesis, tissue culture, enzyme engineering, alcohol fermentation and immuno-technology as areas of immediate interest.

BIOTECHNOLOGY FOR THE COTTON FARMER PROBLEMS AND PROSPECTS

Cotton or white gold as it is aptly called is grown for its lint and seed which yield cotton fibre and seed oil, respectively. This crop occupies 32-33 mha of world area with a production of 20-25 metric tonnes. In India its area spans over 7.5 mha with an average yield of 290 kg/ha of lint and 870 kg/ha of seed cottom.

To meet the challenges of 2000AD with a population of more that 960 million, a total production of 19 million bales is required as against the 13-14 million bales of today. This can be achieved by the use of improved crop production practices coupled with appropriate pest management tactics. In addition, generation of novel. Transgenics may help achieve the near impossible.

Genes that have been identified as potentially profitable, if engineered into acceptable cultivars can be used to generate such transgenics. Among these are genes imparting resistance to herbicides, insects, pathogens and abiotic stresses. It is also widely accepted now that a number of other qualitative characters can be improved, such as fibre strength, fineness, colour and thermal adaptability of the fibre.

Thansgenic plants have become realistic components of stress management world over.Bollworm and herbicide resistant transgenic cotton have received the approval of the Environmental Protection Agency and have been commercially released in the US for cultivation.

Considering the fact that numerous biotic and abiotic stresses limit cotton production, it is likely that future strategies might orient towards the development of a multiadversity resistant high yielding transgenic cotton variety with superior fibre qualities. The most important aspects in the development of transgenic plants are:

- Identification, isolation and clining of the desirable genes.
- Transformation of plant cells or tissues with a suitable vector.
- Regeneration of the transformed cells or tissue.
- Confirmation of the gene integration and expression.

- Hardening and field establishment of the transgenics.
- Major achievements worldover in the field of cotton transgenics.

GENES

Genes for Insect Resistance Used

- Toxin genes from Bacillus thuringiensis.
- Protease inhibitor genes from plants & the horn worm.

Genes used for herbicide resistance :

- Nitrilase gene from a bacteria Klebsiella spp. for Bromoxynil resistance.
- Mono-oxygenase gene from a bacteria Alcaligenes eutrophus for 2,4-D resistance.
- Acstolactate synthase gene for resistance against sufonyl urea & imidazoline group.
- ERPSP synthase gene for glyphosate resistance.

Genes used to Improve fibre qualities :

- Polyhydroxybutyrate & polyhydroxyacetate from bacteria.
- Expression of indigo pigment in fibres.

Genes to be Used

- Cholesterol oxidase genes against insects.
- Heliothis stunt virus genes against Heliothis.
- Genes encoding chitinases glucanases, attacins cecropins vital coat proteins etc. against diseases.

Transformation

- Agrobacterium medicated transformation.
- Particle acceleration gene delivery systems have been successfully used all over the world.

Regeneration

- Callus fissue regeneration-mostly genotype specific.
- Protoplasts were regenerated successfully.
- Meristematic tissues were successfully regenerated.

Expression

- Enhanced expression was achieved using truncated forms of the full length genes, repeat copies of the propoter, modification of the bacterial codons to plant prefered type.

Commercial Release:

- Bollworm resistant transgenics-NUCOTN 33M & NUCOTN 35B were released in the US and cotton transgenics-are in now under field cultivation in Australia, South Africa and China.

- 8 lakh hectares were planted with the transgenic cottons in 1996 and 1997.
- More than 8,000 hectares were damaged by bollworms.
- Transgenic cotton plants in Australia were also found to be attacked by the bollworms.
- Herbicide (Bromoxynil) resistant transgenics BXN 57 and BXN 53 were released in the US in 1996.

Problems Encountered/Expected

- Gene silencing and expression instability.
- Resistance in insects to the toxins used. Studies on baseline toxicity indicated that there is a natural variation in the ability of Helicoverpa armigera in tolerating the CrylAc toxins. Some strains are capable of surviving the totoxins by virtue of an in built capacity to tolerate the toxins, while others are not. While development of Helicoverpa resistance to the CrylAc toxins itself may not be of immediate concern, a slight shift in the tolerance levels resulting from continuous exposure to the transgenics, can cause control difficultes.
- High cost of transgenic seed.
- Microecological changes resulting in pest shifts. CrylAc is a broad spectrum lepidopteran toxins that can cause significant changes in almost all populations of lepidopteran insects occurring on the cotton crop. Some of these insects, which cause negligible economic damage to the crop, harbour parasitoids which have the potential to keep the bollworm populations under check. Consequently, in the absence or low populaions of natural enemies, the bollworms can emerge as stronger pests that before. Insects pests such as Spodoptera litura which are less affected by the CrylAc toxins and which have been under check due to the use of pyrethroids can resurgace as major pests as pyrethroid use is likely to be reduced on transgenic Bt cottons.

Some Recent Development in Cotton Transgenics

- Three Bt transgenic lines, 95-1, RIOI and R-19 have been indigenously developed in China through Agrobacterium medicated (Somatic embryo) and pollen tube pathway transformation. Thestability of Bt gene expression and field efficacy have been confirmed.
- Microprojectile bombardment parameters for pollen mediated transformation and genetype independent protocols have been develop in China.
- A new methods of transformation through injection of Agrobacterium into developing embryos, has been reported from South Africa.

- Stable transformation and regermation has been reported for Bt transgenics in Uzbekistan, China, Egypt and Australia. Pakistan reports development of cotton transgenic plants resistant torearcurl virus.
- An antisense DNA of CLCUV DNA-A bome ACI gone along with the antesense DNA of the AC2 and AC3 gene was used for the vector construction and transgenic cotton resistant to the CLCUV was developed in Pakistan.
- In Russia, A glucanase gene has been isolated from thermo philic bacteria and is being used as a new reporter system fortransgenic cotton development.
- Genetically engineered Baculovirus ACMNPV with gencs from straw itch mite has been developed been developed by Monsanto and Zeneca with researchers from Madison, USA.
- Bt rransgenic cotton currently occupies 20% (85,000 ha) of cotton area in Australia. Isecticide use has been reduced by about 60% The efficacy has been moderate. The transgenic plants were found to express Bt proteimns only till the 95th day after which fluctuaions were observed. The reduction in expression was primarily due to down regulation which post transcriptional changes of the unstable RNA It was also reported that cotton tannings which increase with growth phase, act as antagonistswith Bt toxins.
- After two years of field cultivation, true resistance to Bt has not yet been detected in Australia.
- Reslistanor to Bt in field populations of Helicoverpa armigera is reported form China
- Annually 100 million dollars are now being spent in the US, only in search of new insecticidal genes.
- In the US, 45% of cotton area this year will be under transgenic cotton (herbicide and insect resistant). However, the area under Bt cotton seems to on the decline.
- Transgenic cotton developed in Australia using the Stunt virus genes, is not performing well.
- Transgenic cotton developed in the US using lectin genes, is not performing well against Heliothis virscens.
- The Australian report that Helicoverpa armigera is at least 100 times less susceptible to Cry toxins as cinoared to Heliothis virescens.
- The Window strategy for insecticide resistance management in Australia is still important for transgenic cotton.
- Cotton Pest spectrum in the US is getting altered after the introduction of Bollgard. Unsprayed Bt cotton sustained 4 times more attack of tarnished bugs, 2.4 times more with boll weevil, 2.8 times more with stink bugs and Spodoptera.

- Due to these changes in pest complex farmers had to spray 3-5 times on boligard as compared to 6-8 times on non-Bt cottons.

PROSPECTS AND POTENTIAL

The application of biotechnology in cotton farming can be either in the form of production of fermentation products or novel recombinant products for use (example, recombinant Pseudomonas expressing CrylAc) or as transgenic plant with in buitt resistance to biotic and abiotic stresses. Transgenic crops with in built resistance to insect pests and diseases can be extremely useful as this would result in the reduction of insecticide use apart from making pest management simple for the farmer.

Introduction of the bollworm resistant transgenic cotton is expected to reduce the use of chemicals used to control bollworm especially Helicoverpa importantly at a time when resistance to most insecticides including pyrethrolids is on the increase all over the world.

Even if the introduction of the Bt cotton in India could result in a 25-30% reduction in insecticide use on cotton, this would mean a benefit of about of about Rs 300 crores, apart from the f-avourable impact on the environment so far, transgenic plants have been produced in about 60 plant species.

Cotton has received special attention of the biotechnological companies in the developed countries who were attracted by the profit motives associated with the high value added to the transgenic seeds.

The firs Bt transgenic cotton has already been released, as Bollgard. The Delta and Pineland using DP 5410 and DP 5690 as recurrent parents. The D and PL brand BT transgenic were labelled as Nucotn 33B Two million acres were planted in 1996 in mid-south region of USA, mainly in regions where Heliothis virescens was problematic. Nearly 1.2 M acres were under Nucotn 33 B and the rest under Nucotn 35B.

The Bollgard brand Bt cotton seed was sold at US $34-36 as compared to US $8-9 per hectare for non-engineered cotton seed. The average cost of control of cotton insect pests in the US was approximately US $150 in the early 1990s. Hence the prices were still found to be attractive. The transgenics were found to be more effective against Helothis virescens as compared to pectinophora gossypiella and Heliothis zea.

Farmers growing transgenics had to sign'an agreement with D & PL stating that they would not keep seed for planting next year. Quick ELISA tests have been devised to test for Bt toxins in plant parts, to check the illegal spread of transgenics.

Two more transgenic cotton varieties tolerant to herbicide Bromoxynil, trademarked as BXN 57 and BXN 53 were developed from the recurrent paren Coker 315 BYTHE Stonevuille pedigree company in collaboration with Calgene Inc. This seed was priced nearly 1.7 times the straight varieties.

In Australia the CSIRO and Narrabri research centre, in collaboration with Monsanto have developed transgenic cotton plarts that were commercial released in 1997. This introduction resulted in 50-60 % reduction in the US $ 93 million spent by farmers each year on insecticides.

Pakistan which uses 90% of its total insecticides on cotton alone and China which is one of the major consumers of insecticides on cotton are reportedly strongly considering the prospects of introduction of Bt transgenic cottons.

Cotton acreage in India is only 5% of the total cropped area, yet it consumes more than 50% of the total insecticides used in the country.

The pattern of usage, however, is not uniformly, spread, for instance in Andhra Pradesh alone, where cotton cultivation is only 0.3% of the total cropped area of the country, accounts for 17% of the pesticide use on cotton in India. Again within the average picture of the state, coastal AP uses 30% more pesticide than the state average.

The approximate estimates of insecticide sales in Guntur district alone, where cotton is grown in 0.15 mHa, is about US $ M. Clearly areas with maximum pecticides use per hectare, necessitated mostly due to bollworms, are likely targeted market niches for the bollworm resistant transgenic cottons. Moreover, since the price of the transgneic seed in the US is nearly four times the high input cotton cultivators of the irrigated belt Growing Bt transgenics and herbicide tolerant transgenics on marginal land is not recommended by the company as these would not show any impact in the absence of strong insect pressure and broad leaved weeds.

In most areas of rainfed cottons the use of insecticides and pest pressures are low hence, it may not be economical in such belts, to encourage the use of transgenics.

Now transgenic production technology for many crops has become an established routine procedure in many countries including India. The research on transgenic cotton in government funded research labs in India is almost in the final stages. Meawhuile Mahyco-Monsanto biotech is ready to market Bt transgenic cotton (bollgard) in India.

The introduction of transgenics through these companies are to be in the form of FL seeds wherein Bt genes are inherited from the transgenic exotic germplasm. Such seeds may also be priced at a premium and it remains to be seen how enthusiastic the response of Indian farmers would be to the expensive input. Transgenic releases from government organisations would definitely enforce a competitive pricing to restrict the monopoly of the private companies.

The transgenic technology is also likely to be extremely beneficial to the private sector. For instance, a transgenic with resistance to a specific herbicide marginalises the use of other herbicides for chemical weed control and the chemical company holding the patent on the herbicide might be prove to be beneficial to the resource rich farmers. Clearly, here the targeted market is primarily high input farmers. The monopoly of transgenics due to patenting rights held by the biotech companies may also lead to high pricing of the seeds.

A Significant socioeconomic issue that arises from the introduction of transgenics into the Indian farming system is that the high priced seeds may benefits the prosperous and large farmers thus providing a negative externality on small and marginal farmers.

On the other had it is also argued that the developments from the application of biotechnology would be beneficial to low input farming practises wherein the cost of chemical inputs can be minimised. It is now only a matter of time before we experience the full social. economic and environmental impact of transgenics in our country.

SCIENTIFIC AND INDUSTRIAL RESOURCES FOR BIOTECHNOLOGY

As far as scientific and technological production in Latin America, university–industry research relations are almost nonexistent. Universities in Latin America are more oriented to learning and teaching than to doing research. One might say that the only practical relation between universities and industry is that of supplying the latter with a qualified labour force for professional and operative work.

For the sake of illustration, let us look briefly at the case of Mexico, where university research is highly concentrated. The National Autonomous University of Mexico (UNAM) alone concentrates 50% of all research funds in the country. Nevertheless, UNAM's research budget represented only 22% of its global budget for 1988. The rest represented wages and other operating expenses.

The Mexican dilemma is that its science and technology system produces so little technology that most of it has to be imported. Still, there are four categories of university–industry links: training of professionals, creation of basic and applied knowledge, provision of technical services, and production of technology.

Part of the explanation for the weakness of university–industry links in Mexico is the near absence of a commercial criterion in research. Thus, the prevalent criterion tends to be defined in epistemological terms, which may be crucial for science but irrelevant for technology. For technology development, what matters is economic convenience, not epistemological relevance.

There are, however, some institutions doing cutting-edge biotechnology research in Mexico. In total, there are about 25 centres with biotechnology projects broadly defined.

If we narrow the definition of biotechnology to the newer techniques, such as those based on recombinant DNA, the generation of hybridomas, cell and tissue culture, and protoplast fusion, then the number of institutions is reduced to only seven. The rest are involved in traditional biotechnology research. Of those seven, three belong to the National Autonomous University of Mexico, two to the National Polytechnical Institute, one to the Autonomous Metropolitan University and another one to Chapingo Autonomous University. All of these universities are located around Mexico City, although three of the top research centres involved are decentralized in neighbouring cities (one in Irapuato and two in Cuernavaca).

The pioneering institution in promoting industry links has been the National Autonomous University of Mexico (UNAM), reputed as the largest university in Latin America.

It created a "Centre for Technological Innovation" in 1984 to market UNAM's science and technology. Although the explicit objective of the new centre is to market UNAM's science and technology, a former director declared to Expansión, the Mexican version of Fortune magazine, that they wish to preserve their identity as an academic institution.

They are not after tailor-made research, therefore, for the specific needs of companies. Rather, companies should find out what UNAM's researchers are doing that might be of interest to them. The director of one of the firms that has supported UNAM's research stressed that business firms should take advantage of Mexican technology, because it is cheaper and should allow them to compete better in international trade.

In sum, Mexico does have several excellent research groups working in the frontiers of biotechnology. But the number of researchers is very small in comparison to those in advanced capitalist countries, both in universities and in firms. In developing countries, there tends to be a very weak if not totally absent link between universities and industry.

With such great differences in the quantity of scientists and the differing patterns of university–industry linkages, it is hard to be very optimistic about the possibilities for developing countries to compete successfully in such a dynamic field as biotechnology on the basis of their own resources.

It is necessary, therefore, to undertake a more realistic analysis regarding the actual scientific potential of developing countries and what public policies it would take to translate research into products that might solve their specific socioeconomic and environmental needs. Unless the proper policy incentives exist, it is unlikely that these needs will be taken into account by the private firms doing biotechnology research, or by scientists from advanced countries who have other concerns as their central motivations.

ECONOMIC SCALE AND BIOTECHNOLOGY

Contrary to what optimistic analysts of biotechnology hold, it seems that the main tendencies indicate a bias favouring the large scale, rather than scale neutrality. That is to say, far from favouring small-scale farm operations, the new biotechnologies will tend to reinforce the bias introduced by modern agricultural technologies by requiring strong capital investments. With the new technologies then, the agricultures of developing countries are likely to require a smaller workforce, even if we assume that many of the products of biotechnology may be adopted by those countries.

The alternative might be that, if new technologies are not adopted, and given the global scope of today's world market, agriculture in developing countries may cease to be a viable economic activity altogether. This would be a major paradox, for about 60% of the population in developing countries depends on agriculture. The following examples of new biotechnology products clearly reflect an economic bias towards large-scale production units. These are the implications of the "technological paradigm" represented by modern agriculture.

The generation and sale of cattle and sheep embryos is generally reserved for large and modern production units, which utilize the most advanced genetic technologies. Commercial cloning of livestock is becoming a reality. There are research groups in the U.S. that are developing cell fusion methods with the potential to create multiple, perhaps unlimited, copies of individual cow-and sheep-embryos.

The target in cloning is to produce genetically predictable animals, with superior milk-and meat-production capabilities. University of Wisconsin Professor Neal First and colleagues have successfully cloned cow embryos, but it will be years before they can determine the quality of the resulting breed. For what they clone is not the mother cow but the embryo itself, *i.e.*, an already fertilized egg. One of the goals of researchers is to improve the cloning technique to develop the entire process in test tubes and eliminate the need for embryos from the cows.

One implication of this type of research is that, in the future, it will be possible to produce "elite" animals for meat and milk production. This genetic improvement of animals, combined with the use of growth hormones, promises to revolutionize production in these fields, and they will probably displace large numbers of farmers, given the large-scale bias. The dictum for this type of modern operation will continue to be, perhaps in an exacerbated manner, "get big, or get out of farming."

Socially, this trend points to a future agriculture that will require not much more than a percent or two of a country's labour force. The main concern will be with how to employ productively, with adequate incomes, those people who are displaced from agriculture. In the U.S., this stage has already been reached: only 2% of the labour force is directly employed in farming.

Another factor that reinforces the previous interpretation regarding the economic bias towards the larger scale is that new agriculture will probably combine two or more products of the current technological revolution, including computers, for the integrated management of agricultural enterprises. It is very likely that a systems approach to management will be reinforced. For instance, and continuing with the example of livestock production, the most sophisticated dairy operations may use embryo transfer technologies to improve the herd, while using computers to maintain exact yield records per cow and to monitor their feed requirements.

The result will be that only large-scale operations will be able to make the heavy capital investments required for such integrated systems for managing resources. All of this reinforces the economic-scale tendency to discriminate against small-scale producers and accelerates the process of "depeasantization without full proletarianization". The dilemma involved in this process is that, although farming expels many of its workers because of increased capital-intensity, the larger economy cannot absorb all those redundant workers. This is the source of the swelling "informal sector" in Latin American economies.

STUDYING BIOTECHNOLOGY' FUTURE IMPACTS

Because agricultural biotechnology products are still largely in a laboratory stage, there is room to hope that some social forces will be able to steer their development in more desirable directions. This calls for enlightened policy by governments and international agencies that must be ahead of the competitive dynamics that drive businesses and technological innovation.

A tough call indeed, for governments must function within the confines of parliamentary committees or government agencies, whereas companies and researchers follow a much swifter market temporality.

It is between the logics of state and markets where sociological analysis can help identify areas of social organization where some policy and action may be effectively directed. A new approach to study the emerging dynamics of global capitalism and suggests ways in which it could be utilized in the assessment of future impacts of biotechnology.

There are actually two approaches that may be combined in studying the future impacts of biotechnology. One covers the global dynamics of capitalism, the "global commodity chains" approach (GCC), and the other might be termed "bottom-up linkages" (BUL) approach.

The GCC approach addresses the links among households, enterprises, and states in global production, whereas the BUL approach highlights the importance of local conditions and social actors in future technological innovation.

With these two approaches one can move beyond and below the analysis of the nation state in studying the course of development. Let us begin with a brief description of the GCC approach. William Friedland (1984) first proposed the "commodity systems" approach to understand the dynamics of global capitalism with regard to agricultural commodities. More recently, the contributors to Gereffi and Korzeniewicz (1994) developed a " global commodity chains" approach that is complementary to that of Friedland.

On the one hand, as Laura Raynolds puts it, Friedlands analysis "focuses on the organization of production (including production scale, labour organization, and the role of science and technology) and its integration into marketing and distribution systems". On the other hand, the commodity chains approach, a term first proposed by Hopkins and Wallerstein (1986):

emphasizes the interlocking "nodes" of production that go into creating a finished commodity. This latter formulation is more sensitive than the former [Friedlands] to the links between component production processes, the geographical location (and potential dispersion) of production, and the variability of "commodity chains" over time. Yet Friedlands approach more carefully avoids the reification of commodity systems and places greater emphasis on microproduction relations.

In the case of future biotechnology products, one initial task would be to identify the key global commodity chains that will be affected to concentrate research on those that are regarded as the most relevant by social and environmental criteria.

A global commodity chain (GCC) "consists of sets of interorganizational networks clustered around one commodity or product, linking households, enterprises, and states to one another within the world-economy."

Furthermore, this type of analysis, shows how "production, distribution, and consumption are shaped by the social relations (including organizations) that characterize the sequential stages of input acquisition, manufacturing, distribution, marketing, and consumption".

Some key concepts of comparative sociology, such as national development and industrialization, are increasingly perceived as problematic with this multilayered approach. For the locus of industrialization may no longer be an advanced capitalist country but a developing country, yet the control over the GCC continues to reside in the wealthier countries whose service industries seem to thrive.

Viewed in a global context, however, it becomes clear that the service sectors in wealthier countries are firmly linked to other productive sectors in a GCC, which shreds the thesis of postindustrial societies to pieces. In other words, manufacturing continues to be at the root of global capitalist dynamics, and the so-called service industries also entail transformation to the extent that they are based on human skilled judgement.

Other important distinctions that have been elaborated by the GCC approach regard the dominant force within each commodity chain. This refers primarily to the type of market in question: is it an oligopolistic or an oligopsonic market, that is, one dominated by a few producers or by a few buyers?

Oligopolistic GCCs are called "producer-driven" chains, whereas those dominated by a few buyers are called "buyer-driven" chains.

This distinction is important in the case of agricultural production, even though most commodity chains are producer driven. In the agricultural inputs sector, for example, the agricultural machinery companies or the petrochemical companies largely determine the technologies supplied for agricultural production. They dominate their backward linkages with their own suppliers and, because the industry is highly concentrated, they also dominate their forward linkages with their consumers, the farmers.

Furthermore, there is the prevalence of contract-farming arrangements by producer firms that also act as a monopsony. For example, Campbells controls about 90% of the canned-soup market in the U.S. and is the single buyer of a number of agricultural commodities, namely tomatoes, from thousands of farmers. Similarly, many agricultural commodities such as many grains tend to be dominated by commercial capital or "buyer-driven" commodity chains. Witness the domination exercised by companies such as Cargill in the soybeans world market.

Whether biotechnology products will be channeled through producer-driven or buyer-driven GCC will likely involve a different combination of policies in trying to shape their social and environmental impact. For instance, Rodolfo Quintero (1992) recently suggested that tomatoes embody such a large relative value and are such an important export crop for Mexico that biotechnology research efforts should focus on this particular commodity.

Although these facts may be true, we should also ask the social question of who controls this commodity chain and to whose benefit would such research efforts accrue. It is not enough to do an economic calculation in terms of value added and trade balances; one also has to study issues of employment, whether small-, medium-, or large-scale farmers dominate within each commodity chain, and the type of environmental practices that prevail.

Questions should also be raised as to the type of GCC that dominates the agricultural inputs sector that supplies a given commodity chain. Will it be possible to establish a new marketing network in a commodity chain with already firmly established agrochemical firms, *e.g.*, Monsanto? Or would it be best to seek an alliance with such companies, trying to reshape their research agendas in more benign directions? With regard to the "bottom-up linkages" approach, the key question is the extent to which local environmental and social conditions are taken into account in technology development.

With conventional agriculture, scientific approaches have come to dominate the generation of technology, largely abandoning local knowledge and practices. Just as much biodiversity has been lost to the planting of homogeneous modern plant varieties, so has much local and indigenous knowledge been wiped out by the sweeping introduction of conventional agriculture in developing countries.

In the case of Latin America, our scientists tend to be trained in Northern countries and they adopt the technological paradigm of conventional agriculture. There has been much disregard, therefore, for building on the local conditions and local knowledge of farmers when introducing technological innovations.

With the hegemony of neo-liberalism and the dominant trend towards the globalization of the world economy, the forces of conventional agriculture can only get stronger.

But public concern for social and environmental sustainability also runs high. It will take democratic politics and decision-making, therefore, to address these concerns beyond the dictates of the market and profitability concerns. Only if public institutions, including universities, and grassroots farmer organizations establish an alliance can there be any hope to move into a more desirable form of development.

THE PATTERN IN DEVELOPED COUNTRIES

In the industrialized countries, a new pattern of biotechnological research funding has emerged. With the availability of property right protection of biotechnologies and prospects of vast markets for biotechnology products and the techniques, the bulk of the research is funded, carried out, and controlled by the private sector.

For instance, since 1976 when GENENTECH, the first biotechnology company, was established, the new techniques have spawned a variety of industries that now comprise more than 400 start-up firms, more than 200 established firms that have diversified into biotechnology, and more than 200 supply firms in the United States alone.

The new biotechnology industry in the United States produced pharmaceutical, diagnostic tests and agricultural products worth close to US$2 billion in 1990. A similar trend is seen in Europe and Japan.

It is estimated that about 60 percent of the biotechnology research and development funding in industrialized countries is from the private sector. Thus, the private sector has been a major force in enhancing the capability of these countries in this field.

Research institutions in the public sector are now generally required to raise a substantial part of their budget from non-government sources, via contractual research, licensing agreements and royalties.

This is tending to increase secrecy over research findings and to hinder free scientific communication. University professors, researchers and government institution scientists are increasingly becoming entrepreneurial and are entering private industry.

Another important trend is that large multinational corporations are purchasing smaller seed and biotechnology companies and diversifying their holdings. This allows them to develop a package sale of chemicals, seeds and equipment. Heavy involvement of the private sector and market considerations greatly influence the topics and commodities chosen for research.

Major crops, commodities and farming systems of great socio-economic importance to the developing world, but of little international market importance, do not figure in the biotechnology research agenda of industrialized countries. Furthermore, these countries are keen to reduce their production costs, increase the productivity, quality and value of their products and, thus, improve their overall competitiveness in the world market.

THE PATTERNS IN DEVELOPING COUNTRIES

Biotechnology facilities are being established in most developing countries. However, the level of research, development and use of biotechnology for agriculture, forestry and fisheries in the developing countries is generally far below the level in the industrialized countries. Among developing countries, the status varies considerably.

A few, such as Brazil, Mexico, India, China and The Republic of Korea, have sought to gain full scientific and technological capacity, especially in agricultural biotechnology. Others, such as Indonesia, Malaysia, the Philippines and Thailand and a few countries of Latin America, have built the capacity to apply biotechniques and develop biotechnologies useful for their agriculture and food industries.

The participation of the private sector in gaining biotechnological capacity is not significant in most of these countries. Many developing countries have inadequate funding, poor human resources and limited access to information, resulting in a relatively low-level capacity for research and technology development and exploitation, especially in the field of modern biotechnology research, which is costly and requires highly trained personnel.

Most developing countries are vague as to their immediate aims in agricultural biotechnology. Few have appropriate proprietary-right protection systems or mechanisms to increase their access to protected techniques and products. Furthermore, there is negligible involvement of the private sector, which accentuates the problem of insufficient attention to biotechnology.

One of the major constraints on biotechnological development in developing countries is the poor quality and extent of higher education in frontier sciences, especially molecular biology. Besides, there are no or very weak links between universities and research institutions to reinforce each other and to utilize synergistically the scarce trained personnel resources. The university experts play a negligible role in national policy formulation, including that on biotechnology. Furthermore the universities are generally not oriented for efficiency in commercialization and are therefore not able to market biotechnology products of their own or others' inventions.

Some international and donor-country public sector institutions strive to ensure that attention is given to solving, through biotechnology, the food and agriculture problems of developing countries. Among this group are the International Agricultural Research Centres (IARCs) located or operating in developing countries.

Capacity exists in several of these centres for utilization of advances in biotechnologies to solve certain problems of plant production and protection and of livestock production and health, and capacity is being built in others. For example, the International Rice Research Institute (IRRI) is incorporating modern molecular biology in its rice research activities.

Its programme is linked closely to the Rockefeller Foundation Biotechnology Network. The World Bank, the International Service for National Agricultural Research (ISNAR) and the Australian Government undertook a major study on the likely impact of modern biotechnology on agriculture.

This study is being followed up by the participants by an expansion in the World Bank lending for biotechnology and by ISNAR's establishing an Intermediary Biotechnology Service to provide advisory services to national agricultural research systems.

APPLICATIONS, IMPACT AND POTENTIAL

Biotechnological research and development are moving at a very fast rate. For instance, three years ago, transgenics in rice, especially in japonica rices, were considered rather difficult because of the problem of regeneration through protoplast culture.

But today, a large number of transgenics both in indica and japonica types are available and are being tested at various stages. It is therefore difficult to predict the potential and impact of a given technology beyond five years or so. The current level of technology and refers to the developments likely to occur in the next three to five years.

The potential of genetic engineering in medicine has received much attention. The potential for advances in agriculture, forestry and fishery are similarly bright. Biotechniques are already being used to create new strains of crop plants, new plant and animal diagnostic products, animal vaccines, biological pesticides and herbicides, other biological control agents, and modifications in domestic animals used for food production.

In some cases, applications are being held in check by the need for still more research to ensure that there are no harmful effects (or, if there are, they are greatly outweighed by the benefits) and by the slow pace of evolution of regulations governing the release of genetically modified organisms or products for general use or acquisition.

Biotechnological research, especially genetic engineering, on problems of field and tree crops, forest species and fisheries has a relatively short history.

The knowledge of biological aspects of species and ecosystems needs to be expanded in may cases before new biotechnologies can be applied effectively. Even where transfer of one or more genes can be achieved reliably, expression of the transferred gene(s) may not occur in the expected way.

The search for and identification and cloning of useful genes, effective vector systems, methods of gene transfer and promoter mechanisms should be intensified.

A perspective on both the difficulties and intensity of the research effort suggests that diagnostic tools will speed solutions in the next five to ten years to certain disease problems in several crop, forest, animal and fish species, such as black sigatoka of banana and plantain, and virus/viroid diseases of coconut and rice.

Improved tolerance to certain physical stresses, such as heat tolerance in potatoes, may also be achieved, but prospects for improved resistance of varieties and clones to most abiotic stresses are longer term.

CROPS

Modern biotechnologies can add greater precision and speed to plant breeding. Transgenics have already been reported in 40 crop plants, including maize, rice, soybean, cotton, rapeseed, potato, sugar beet, tomato, potato and alfalfa, but the new varieties are yet to be used commercially.

Near-future opportunities for commercial exploitation include vegetables and fruits (potato, tomato, cucumber, cantaloupe and squash), followed by legumes (alfalfa) and oilseed crops (oilseed rape).

A good number of the transgenics are herbicide-resistant plants whose widespread use is somewhat controversial. Wide use is currently made of tissue culture techniques for micropropagation of elite clones and for freeing planting materials of pathogens. Monoclonal antibodies are also in use as diagnostic aids in the detection and identification of viruses and viroids.

Another culture and microspore culture giving rise to haploids are being used in variety improvement to facilitate and accelerate breeding. Molecular maps and markers are being widely used to identify genes of interest to accelerate conventional breeding programmes.

Efficient biological nitrogen fixation systems and strains for efficient utilization of soil nutrients are being genetically engineered. Other long-term objectives are the genetic manipulations of photosynthesis pattern and the production of hybrid seed through apomixis. A very distant possibility is providing nitrogen fixation capacity to cereals.

LIVESTOCK

Among agricultural and allied fields, animal production and health have benefited the most from biotechnology, although practical use of transgenic livestock is only a future possibility. Wide use of monoclonal antibodies for efficient diagnostics, leading to safe and specific treatments of animal diseases, is a major breakthrough.

Through genetic engineering, vaccines for the prevention of viral, bacterial, and parasitic animal diseases are rendered more effective and safe. Tailored vaccines exist for pig scours, chicken bursar disease and cattle tickborne diseases. Pathogen-specific vaccines are attractive goals.

Other interesting possibilities are endocrine-directed vaccines to stimulate twinning in beef cattle, immunocastration, livestock growth rate increase and vaccines that compensate for various stress-induced production losses.

Advances in genetic engineering will also facilitate the production of male-only populations of screwworms, tsetse flies, ticks and various other ectoparasites for use in the sterile release technique of control or eradication. Furthermore mammalian tissue culture may replace whole animals in the 1990s for toxicity testing of certain chemicals.

The culture technique can also be exploited for studying and analyzing pesticide metabolism and for herbicide pre-screening. In vitro fertilization and embryo sexing techniques have considerably increased the use of embryo transfer techniques for cattle breeding and trade.

The value of the approach will be further enhanced if embryo cloning techniques can be reliably employed. Microbial and enzymatic treatment of roughages and genetic engineering of rumen bacteria both have great potential to improve animal nutrition.

Growth hormones can be produced by genetically engineered microorganisms in quantities and at the low cost necessary for widespread use to speed and increase milk and lean meat production. Biotechnological tools (embryo culture, gene cloning, etc.) may also be used for conservation of genetic resources.

FORESTRY

Successful regeneration through micropropagation or somatic embryogenesis has been achieved for about 100 forest species although, for most, considerable development work would be required before commercial propagation could be contemplated.

In breeding programmes incorporating clonal testing, inclusion of a micromultiplication phase may facilitate more rapid deployment of superior genotypes than that afforded by sequential multiplication by cuttings.

This will particularly be the case when gene mappling techniques are sufficiently advanced to permit accurate identification of superior genotypes without replicated clonal testing in the field, especially within a breeding line where substantial genetic disequilibrium exists. Furthermore, gene mapping may allow eventual identification of valuable genes, including those that contribute to quantitative traits. As regards the application of genetic engineering, many major traits of commercial importance to forestry are under polygenic control, and much remains to be learned about the operation of the genes involved before this technique can have a major impact for these characters.

An earlier application of genetic engineering in some forestry operations may be the introduction of genes, some already available, known to confer insect of disease resistance. Other applications of biotechnology with obvious value for forestry, but as yet not widely supported by experimental successes, include: in vitro manipulation of the maturation state, *e.g.* the promotion of early flowering to reduce generation intervals, in vitro selection for traits such as disease resistance and tolerance to salinity; and the use of haploid cultures.

FISHERIES

Two broad areas of biotechnology exist within the fisheries sector, namely natural products biotechnology (including mostly marine species); and aquaculture biotechnology.

Commercially valuable products such as pharmaceutical compounds, pigments, oils, sterols, alginates and agarose are being extracted from micro- and macro-algae in many parts of the world. Aquatic invertebrates are currently being screened for biologically active compounds that may have, inter alia anti-carcinogenic or antiviral properties, whereas marine bacteria are currently utilized in treatments of oil spills and holding tanks in tanker ships.

Within aquaculture, induced increases in the chromosome complement (polyploidy) of commercially important species such as salmon and oysters have increased the growth and marketability of these species.

Genetically engineered microbes can be used to produce fish growth hormones, which might then be used to improve feed conversion and growth rate. Synthetic reproductive hormones are produced commercially and used to regulate fecundity, breeding cycles, growth rates and sex determination in certain cultured species.

In attempts to increase desirable culture qualities such as growth rate, disease resistance, temperature tolerance and marketability, transgenic fishes containing introduced genes from other species have been produced on an experimental scale. The application of advanced biotechnology to the fisheries sector is a relatively recent practice. Therefore, continued research can be expected to reveal additional commercially viable applications and products within this sector.

2

Benefits of Cover Crops and Green Manures

CROP ROTATION WITH COVER CROPS

One of the biggest challenges of cover cropping is to fit cover crops into your current rotations, or to develop new rotations that take full advantage of their benefits.

Some of the systems used successfully by farmers in different regions of the U.S. One might be easily adapted to fit your existing crops, equipment and management. Other examples may point out ways that you can modify your rotation to make the addition of cover crops more profitable and practical.

Whether you add cover crops to your existing rotations or totally revamp your farming system, you should devote as much planning and attention to your cover crops as you do to your cash crops.

Failure to do so can lead to failure of the cover crop and cause problems in other parts of your system. Also remember that there is likely no single cover crop that is right for your farm.

Before you start:

- Review Benefits of Cover Crops and Selecting the Best Cover Crops for Your Farm.
- Decide which benefits are most important to you.
- Then consider how these cover crop rotations might be adapted to your particular conditions.
- Talk to your neighbours and the other "experts" in your area, including the contact people listed in Regional Experts.
- Start small on an easily accessible plot that you will see often.

Be an opportunist—and an optimist. If your cropping plans for a field are disrupted by weather or other conditions outside of your control, this may be the ideal window for establishing a cover crop. Consider using an early-maturing cash crop to allow for timely planting of the cover crop. Cover crops can be used for feed. Consider harvesting or grazing for forage or alternative livestock such as sheep and goats.

COVER CROPS FOR CORN BELT GRAIN AND OILSEED PRODUCTION

In addition to providing winter cover and building soil structure, nitrogen (N) management will probably be a major factor in your cover crop decisions for the corn>soybean rotation.

A fall- planted grass or small grain will scavenge leftover N from the previous corn or soybean crop. Legumes are much less efficient at scavenging N, but will add N to the system for the following crop. Legume/grass mixtures are quite good at both.

Corn>Soybean Systems

Keep in mind that corn is a heavy N feeder, soybeans benefit little, if at all, from cover crop N and that you have a shorter time for spring cover crop growth before corn than before soybeans.

Precaution: If you use herbicides, be sure to check labels for plant-back or rotation intervals to ensure that your cover crop isn't adversely affected.

Cover crop features: Rye provides winter cover, scavenges N after corn, becomes a long-lasting (6 to 12 week) residue to hold moisture and suppress weeds for your soybeans; hairy vetch provides spring ground cover, abundant N and a moderate-term (4 to 8 week) mulch for a no-till corn crop; field peas are similar to vetch, but residue breaks down faster; red clover is also similar, but produces slightly less N and has less vigorous spring regrowth; berseem clover grows quickly to provide several cuttings for high-N green manure. Here are some options to consider adapting to your system:

Corn>Rye>Soybeans>Hairy Vetch

In Zone 7 and warmer, you can grow a cover crop every year between your corn and full-season beans. Also, you can use wheat or another small grain to replace the cover crop before beans, in a three-crop, two-year rotation (corn>wheat>doublecrop beans).

In all cases, another legume or a grass/legume mixture can be used instead of a single species cover crop. Where it is adapted, you can use crimson clover or a crimson/grass mixture instead of vetch.

In cooler areas, plant rye as soon as possible after corn harvest. If you need more time in the fall, try overseeding in rowed beans at drydown "yellow leaf" stage in early fall, or in early summer at the last cultivation of corn. Seeding options include aerial application where the service is economical, using a specialty high-body tractor with narrow tires, or attaching a broadcast seeder, air seeder or seed boxes to a cultivator.

Precaution: Broadcast seeding of cover crops into standing crops is less dependable than other seeding methods. Success will depend on many factors, including adequate rainfall amount and distribution after seeding.

Kill the rye once it is about knee-high, or let it go a bit longer, killing it a couple of weeks before planting beans. Killing the rye with herbicides and no-tilling beans in narrow rows allows more time for cover crop growth, since you don't have to work the ground. If soil moisture is low, consider killing the rye earlier. Follow the beans with hairy vetch or a vetch/small grain mixture.

Legumes must be seeded at least 6 weeks before hard frost to ensure winter survival. Seed by drilling after soybean harvest, or by overseeding before leaf drop. Allow the vetch (or mixture) to grow as long as possible in spring for maximum N fixation.

Harvesting sweet corn, seed corn or silage corn opens a window for timely cover crop planting. Harvest or graze the small grain or legume/small grain mixture in spring if needed for feed.

In Pennsylvania, Ed Quigley seeds rye or spring oats after corn silage harvest. The oats can be cut for silage in fall if planted by early September. Rye can be made into rylage or sprayed before no-till corn the following spring.

Worried about planting your corn a bit late because you're waiting for your cover crop to mature? Research in Maryland, Illinois and elsewhere suggests that no-tilling corn towards the end of the usual window when using a legume cover crop has its rewards.

The delay can result in greater yields than earlier planting, due to greater moisture conservation and more N produced by the cover crop, or due to the timing of summer drought. In Pennsylvania, however, delayed planting sometimes reduced corn yields following rye.

Check your state variety trial data for a shorter season corn hybrid that yields nearly as well as slightly longer season corn. The cover crop benefit should overcome many yield differences.

Worried about soil moisture? There's no question that growing cover crops may consume soil moisture needed by the next crop. In humid regions, this is a problem only in an unusually dry spring.

Time permitting, allow 2 to 3 weeks after killing the cover crop to alleviate this problem. While spring rainfall may compensate for the moisture demand of most cover crops by normal planting dates, cover crops can quickly dry out a field. Later in the season, killed cover crop residues in conservation tillage systems can conserve moisture and increase yields.

In any system where you are using accumulated soil moisture to grow your cash crop, you need to be extra careful. However, farmers and researchers are finding that water-thrifty cover crops may be able to replace even a fallow year without adversely affecting the cash crop.Corn>Rye>Soybeans>Small Grain>Hairy Vetch. This rotation is similar to the corn>rye>soybeans rotation, except you add a year of small grains following the beans. In crop rotation research from different areas, many benefits accrue as the rotation becomes longer. This is because weed, disease and insect pest problems generally decrease with an increase in years between repeat plantings of the same crop.

Residue from small grains provides good organic matter for soil building, and in the case of winter grains, the plants help to prevent erosion over winter after soybeans loosen up the soil.

The length of the growing season will determine how you fit in cover crops after full-season soybeans in the rotation. Consider using a short- season bean if needed in order to achieve timely planting after soybean harvest.

Calculate whether cover crop benefits will compensate for a possible yield loss on the shorter season beans. If there is not enough time to seed a legume after harvest, use a small grain rather than no cover crop at all.

The small grain scavenges leftover N following beans. Legume cover crops reduce fertilizer N needed by corn, a heavy N feeder. If you cannot seed the legume at least six weeks before a hard frost, consider overseeding before leaf drop or at last cultivation.

Precaution: Because hairy vetch is hard seeded, it will volunteer in subsequent small grain crops.

An alternate rotation for the lower mid-South is corn>crimson clover (allowed to go to seed) > soybeans > crimson clover (reseeded) > corn. Allow the crimson clover to go to seed before planting beans. The clover germinates in late summer under the beans.

Kill the cover crop before corn the next spring. If possible, choose a different cover crop following the corn this time to avoid potential pest and disease problems with the crimson clover.

Precaution: In selecting a cover crop to interseed, do not jeopardize your cash crop if soil moisture is usually limiting during the rest of the corn season! Banding cover crop seed in row middles by using insecticide boxes or other devices can reduce cover crop competition with the cash crop.

3 Year: Corn>Soybean>Wheat/Red Clover. This well-tested Wisconsin sequence provides N for corn as well as weed suppression and natural control of disease and insect pests. It was more profitable in recent years as the cost of synthetic N increased. Corn benefits from legume-fixed N, and from the improved cation exchange capacity in the soil that comes with increasing organic matter levels.

Growers in the upper Midwest can add a small grain to their corn>bean rotation. The small grain, seeded after soybeans, can be used as a cover crop, or it can be grown to maturity for grain.

When growing wheat or oats for grain, frost-seed red clover or sweet-clover in March, harvest the grain, then let the clover grow until it goes dormant in late fall. Follow with corn the next spring. Some secondary tillage can be done in the fall, if conditions allow. One option is to attach sweeps to your chisel plow and run them about 2 inches deep, cutting the clover crowns.

Alternatively, grow the small grain to maturity, harvest, then immediately plant a legume cover crop such as hairy vetch or berseem clover in July or August.

Soil moisture is critical for quick germination and good growth before frost. For much of the northern U.S., there is not time to plant a legume after soybean harvest, unless it can be seeded aerially or at the last cultivation. If growing spring grains, seed red clover or sweet-clover directly with the small grain.

An Iowa study compared no-till and conventional tillage corn>soybean>wheat/clover rotation with annual applications of composted swine manure.

Berseem clover or red clover was frostseeded into wheat in March. Corn and soybean yields were lower in no-till plots the first year, while wheat yield was not affected by tillage. With yearly application of composted swine manure, however, yield of both corn and soybean were the same for both systems beginning in year two of the 4-year study.

Adding a small grain to the corn>soybean rotation helps control white mold on soybeans, since two years out of beans are needed to reduce pathogen populations. Using a grain/legume mix will scavenge available N from the bean crop, hold soil over winter and begin fixing N for the corn. Clovers or vetch can be harvested for seed, and red or yellow clover can be left for the second year as a green manure crop.

Using a spring seeding of oats and berseem clover has proved effective on Iowa farms that also have livestock. The mix tends to favour oat grain production in dry years and berseem production in wetter years. Either way the mixture provides biomass to increase organic matter and build soil.

You can clip the berseem several times before flowering for green manure. Precaution. Planting hairy vetch with small grains may make it difficult to harvest a clean grain crop. Instead, seed vetch after small grain harvest. Be sure to watch for volunteer vetch in subsequent small grain crops. It is easily controlled with herbicides but will result in significant penalties if found in wheat grain at the elevator.

VEGETATION MANAGEMENT TO CREATE A COVER CROP MULCH

Herbicides are the most commonly used tools for cover crop suppression in conservation tillage systems. Non-chemical methods include propane flamers, mowing and mechanical tillage.

Mowing a rye cover crop when it heads out in late spring provides sufficient kill. The rye must be in the pollination phase, or later, to be successfully killed. When the anthers are fully extended and you can thump the stalk and pollen falls down, it is time to mow. If mowed earlier, it just grows back. Flail mowers generally produce more uniformly distributed mulch than do rotary cutters, which tend to windrow the mulch to one side of the mower.

Sickle bar mowers create fairly uniform mulch, but the unchopped rye stalks can be more difficult to plant into. If late spring weather continues cool and wet, more rye regrowth will occur than if the weather remains warm and drier. Typically, if rye is mowed at the pollination stage, regrowth is minimal and not a problem to crops grown in the mowed mulch.

In a Mississippi study, flail mowing, or rolling with rolling disk colters spaced at 4 inches, was usually as effective as herbicides in killing hairy vetch, crimson clover and subterranean clover.

Timing is a key factor when using mowing or rolling to control cover crops. Mechanical control was most effective when the legumes were in the seed formation growth phase (mid to late April) or when stem lengths along the ground exceeded 10 inches. If mowing was followed with a pre-plant herbicide application of Atrazine, the legume kill was even more effective.

Researchers at Ohio State University developed a mechanical cover crop killing tool used to take out a cover crop without herbicides. They call it an undercutter because it uses wide V-blades which run just under the soil surface to cut off the cover crop from its roots.

The blades are pitched to 15 degrees, allowing the blades to penetrate the soil and provide a slight lifting action. Mounted on the same toolbar behind the cutter blades is a rolling basket to flatten and distribute the undercut cover crop.

The undercutter was tried on several cover crops and effectively killed crimson clover, hairy vetch, rye, and barley. These undercutters could be made from locally available stock by innovative tinkerers. Steve Groff of Cedar Meadow Farm in Lancaster County, Pennsylvania, uses a 10-foot Buffalo rolling stalk chopper from Fleischer Manufacturing. Under the hitch-mounted frame, the stalk chopper has two sets of rollers running in tandem.

These rollers can be adjusted for light or aggressive action and set for continuous coverage. Steve says the machine can be run up to 8 miles an hour and does a good job of killing the cover crop and pushing it right down on the soil. It can also be used to flatten down other crop residues after harvest. Groff improved his chopper by adding independent linkages and springs to each roller. This modification makes each unit more flexible to allow continuous use over uneven terrain. Following his chopper, Groff transplants vegetable seedlings into the killed mulch. He direct-seeds sweet corn and snap beans into the mulch.

Two USDA-ARS researchers, Drs. Aref Abdule-Bake and John Teasdale of the Beltsville Maryland Research Centre, have developed a cover-crop roller that acts, in principle, similarly to Steve Groff's rolling chopper. In their extensive research trials using hairy vetch, they no-till planted tomatoes into a mechanically killed hairy vetch cover crop. Details of their research—and other useful information on flail-mowing of cover crops and direct no-till seeding of sweet corn and snap beans into mechanically killed cover crops. As of this writing the bulletin is available only online because the first printing of it was all distributed to farmers and their advisors in a very short time.

Planting cover crops known to readily winter-kill is another non-chemical means of vegetation management. Spring oats, buckwheat, and sorghum fill this need. They are fall-planted early enough to accumulate some top growth before freezing temperatures kill them.

In some locations, oats will not be completely killed and some plants will regrow in the spring. Winter-killed cover crops provide a dead mulch through the winter months instead of green cover. They are used primarily in regions where precipitation is limited, such as West Texas, and in situations where farmers want to plant early in the spring and avoid over wintered cover crops altogether.

COVER CROPS IN ROTATION

Cover crops can fit well into many different cropping systems during periods of the year when no cash crop is being grown. Even the simplest corn/soybean rotation can accommodate a rye cover crop following corn, which will scavenge residual nitrogen and provide ground cover in the fall and winter.

When spring-killed as a no-till mulch, the rye provides a water-conserving mulch and suppresses early-season weeds for the following soybean crop. Hairy vetch can be planted behind soybeans to provide nitrogen for corn the following spring.

Hairy vetch is *not* a good cover crop to use when small grains are included in the rotation—if the vetch ever goes to seed it can become a terrible weed in the small grain crop. In these cases, crimson clover, sweet clover, or red clover should be used, depending on location.

Many vegetable rotations can accommodate cover crops as well. Buckwheat can follow lettuce and still be tilled down in time for fall broccoli. Hairy vetch works well with tomatoes and other warm-season vegetables. The vetch can be killed by flail mowing and tomato sets planted into the mulch.

Limitations of Cover Crops

The recognized benefits of green manuring and cover cropping—soil cover, improved soil structure, nitrogen from legumes—need to be evaluated in terms of cash returns to the farm as well as the long-term value of sustained soil health.

For the immediate growing season, seed and establishment costs need to be weighed against reduced nitrogen fertilizer requirements and the effect on cash crop yields. Water consumption by green manure crops is a concern and is pronounced in areas with less than 30 inches of precipitation per year. Still, even in the fallow regions of the Great Plains and Pacific Northwest, several native and adapted legumes (such as black medic) seem to have potential for replacing cultivation or herbicides in summer fallow.

There is always additional management required when cover crops of any sort are added to a rotation. Turning green manures under or suppressing cover crops requires additional time and expense, compared to having no cover crop at all.

Insect communities associated with cover crops work to the farmer's advantage in some crops and create a disadvantage in others. For example, certain living mulches enhance the biological control of insect pests of summer vegetable crops and pecan orchards by providing favourable habitats for beneficial insects.

On the negative side, winter legumes that harbour catfacing insects such as the tarnished plant bug, stink bug, and plum curculio can pose problems for apple or peach orchardists in the eastern U.S. Nematodes encouraged by certain legumes on sandy soils are another concern of farmers, as are cutworms in rotations following grain or grass crops.

PEST MANAGEMENT BENEFITS OF COVER CROPS

In addition to the soil improving benefits, cover crops can also enhance many pest management programmes. Ecologists tell us that stable natural systems are typically diverse, containing many different types of plants, arthropods, mammals, birds, and microorganisms. Growing cover crops adds diversity to a cropping system. In stable systems, serious pest outbreaks are rare, because natural controls exist to automatically bring populations back into balance.

Farmers and researchers in several locations have observed and documented increased beneficial insect numbers associated with cover crops. The cover crops provide pollen, nectar, and a physical location for beneficial insects to live while they search for pest insects. Conservation tillage proves a better option than tilling because it leaves more crop residue on the surface to harbour the beneficial insects.

Strip tilling or no-tillage disturbs a minimum of the existing cover crop that harbours beneficial insects. Cover crops left on the surface may be living or in the process of dying. At either of these stages they protect beneficials. Once the main crop is growing, the beneficials move onto it.

By having the cover crop in place early in the growing season, the population of beneficials is much higher sooner in the growing season than would be the case if only the main crop were serving as habitat for the beneficials. Innovative farmers are paving the way by interplanting cover crops with the main crop and realizing pest management benefits as a result. Georgia cotton farmers Wayne Parramore and sons reduced their insecticide and fertilizer use by growing a lupine cover crop ahead of their spring-planted cotton.

They started experimenting with lupines on 100 acres in 1993 and by 1995 were growing 1,100 acres of lupines. Ground preparation for cotton planting is begun about 10 days prior to planting by tilling 14-inch wide strips into the lupines.

Herbicides are applied to the strips at that time and row middles remain untouched. The remaining lupines provide beneficial insect habitat and also serve as a smother crop to curtail weeds and grasses. The lupines in the row middles can be tilled in later in the season to release more legume nitrogen.

Dr. Sharad Phatak of the University of Georgia has been working with cotton growers in Georgia testing a strip cropping method using winter annual cover crops. Planting cotton into strip-killed crimson clover improves soil health, cuts tillage costs, and allows him to grow cotton without any insecticides and only 30 pounds of nitrogen fertilizer.

Working with Phatak, farmer Benny Johnson reportedly saved at least $120/ acre on his 16-acre test plot with the clover system. There were no insect problems in the test plot, while beet armyworms and whiteflies infested nearby cotton and required 8 to 12 sprays to control. Cotton intercropped with crimson clover yielded more than 3 bales per acre compared with 1.2 bales per acre in the rest of the field. Boll counts were 30 per plant with crimson clover and 11 without it. Phatak identified up to 15 different kinds of beneficial insects in these strip-planted plots.

Phatak finds that planting crimson clover seed at 15 pounds per acre in the fall produces around 60 pounds of nitrogen per acre by spring. By late spring, beneficial insects are active in the clover. At that time, 6- to 12-inch planting strips of clover are killed with Roundup herbicide.

Fifteen to 20 days later the strips are lightly tilled and cotton is planted. The clover in the row-middles is left growing to maintain beneficial insect habitat. When the clover is past the bloom stage and less desirable as a preferred habitat, beneficials move onto the cotton. Even early-season thrips, which can be a problem following cover crops, are limited or prevented by beneficial insects in this system. Movement of the beneficials coincides with a period when cotton is most vulnerable to insect pests. Following cotton defoliation in the early fall, the beneficials hibernate in adjacent non-crop areas.

Phatak points out that switching to a whole-farm focus while reducing off-farm inputs is not simple. It requires planning, management, and several years to implement on a large scale. It is likewise important to increase and maintain organic matter, which stimulates beneficial soil microorganisms. Eventually a "living soil" will help keep harmful nematodes and soil-borne fungi under control.

ORGANIC MATTER AND SOIL STRUCTURE

A major benefit obtained from green manures is the addition of organic matter to the soil. During the breakdown of organic matter by microorganisms, compounds are formed that are resistant to decomposition—such as gums, waxes, and resins.

These compounds—and the mycelia, mucus, and slime produced by the microorganisms—help bind together soil particles as granules, or aggregates. A well-aggregated soil tills easily, is well aerated, and has a high water infiltration rate.

Increased levels of organic matter also influence soil humus. Humus—the substance that results as the end product of the decay of plant and animal materials in the soil—provides a wide range of benefits to crop production. Sod-forming grass or grass-legume mixtures are important in crop rotations because they help replenish organic matter lost during annual cultivation.

However, several years of sod production are sometimes required before measurable changes in humus levels occur. In comparison, annual green manures have a negligible effect on humus levels, because tillage and cultivation are conducted each year.

They do replenish the supply of active, rapidly decomposing organic matter. The contribution of organic matter to the soil from a green manure crop is comparable to the addition of 9 to 13 tons per acre of farmyard manure or 1.8 to 2.2 tons dry matter per acre.

Nitrogen Production

Nitrogen production from legumes is a key benefit of growing cover crops and green manures. Nitrogen accumulations by leguminous cover crops range from 40 to 200 lbs. of nitrogen per acre. The amount of nitrogen available from legumes depends on the species of legume grown, the total biomass produced, and the percentage of nitrogen in the plant tissue.

Cultural and environmental conditions that limit legume growth—such as a delayed planting date, poor stand establishment, and drought—will reduce the amount of nitrogen produced. Conditions that encourage good nitrogen production include getting a good stand, optimum soil nutrient levels and soil pH, good nodulation, and adequate soil moisture.

The portion of green-manure nitrogen available to a following crop is usually about 40% to 60% of the total amount contained in the legume. For example, a hairy vetch crop that accumulated 180 lbs. N per acre prior to plowing down will contribute approximately 90 lbs. N per acre to the succeeding grain or vegetable crop.

Dr. Greg Hoyt, an agronomist at North Carolina State University, has estimated that 40% of plant tissue nitrogen becomes available the first year following a cover crop that is chemically killed and used as a no-till mulch. He estimates that 60% of the tissue N is released when the cover crop is incorporated as a green manure rather than left on the surface as a mulch.

Lesser amounts are available for the second or third crop following a legume, but increased yields are apparent for two to three growing seasons. To determine how much nitrogen is contained in a cover crop, an estimate is needed of the yield of above-ground herbage and its percentage of nitrogen.

A procedure to make this determination is available in the *Northeast Cover Crop Handbook*, in *Farmer's Fertilizer Handbook*, and in *Managing Cover Crops Profitably*. A description of these publications complete with ordering information. The procedure involves taking a field sample, drying it, weighing it, and sending a sample off for forage analysis, which includes an estimate of protein content. Once the protein content is known, simply divide it by 6.25 to obtain the percentage of nitrogen contained in the cover crop tissue.

Nutrient Enhancement

In addition to nitrogen from legumes, cover crops help recycle other nutrients on the farm. Nitrogen (N), phosphorous (P), potassium (K), calcium (Ca), magnesium (Mg), sulfur (S), and other nutrients are accumulated by cover crops during a growing season.

When the green manure is incorporated, or laid down as no-till mulch, these plant-essential nutrients become slowly available during decomposition. Dr. Greg Hoyt developed a method for estimating nutrient accruement by cover crops in order to reduce the soil test recommendation of fertilizer for the following crop.

Rooting Action

The extensive root systems of some cover crops are highly effective in loosening and aerating the soil. In Australian wheat experiments, the taproots of a blue lupine cover crop performed like a "biological plow" in penetrating compacted soils. When cover crops are planted after a subsoiling treatment, they help extend the soil-loosening effects of the deep tillage treatment. The rooting depths of several green manures grown under average conditions are listed in.

Weed Suppression

Weeds flourish on bare soil. Cover crops take up space and light, thereby shading the soil and reducing the opportunity for weeds to establish themselves. The soil-loosening effect of deep-rooting green manures also reduces weed populations that thrive in compacted soils.

The primary purpose of a non-legume green manure—such as rye, millet, or sudangrass—is to provide weed control, add organic matter, and improve soil tilth. They do not produce nitrogen. Thus, whenever possible, annual grain or vegetable crops should follow a legume green manure to derive the benefit of farm-produced nitrogen.

Providing weed suppression through the use of allelopathic cover crops and living mulches has become an important method of weed control in sustainable agriculture. Allelopathic plants are those that inhibit or slow the growth of other nearby plants by releasing natural toxins, or "allelochemicals." Cover crop plants that exhibit allelopathy include the small grains like rye and summer annual forages related to sorghum and sudangrass. The mulch that results from mowing or chemically killing allelopathic cover crops can provide significant weed control in no-till cropping systems. Living mulches suppress weeds during the growing season by competing with them for light, moisture, and nutrients.

Soil Microbial Activity

A rapid increase in soil microorganisms occurs after a young, relatively lush green manure crop is incorporated into the soil. The soil microbes multiply to attack the freshly incorporated plant material. During microbial breakdown, nutrients held within the plant tissues are released and made available to the following crop.

Factors that influence the ability of microorganisms to break down organic matter include soil temperature, soil moisture, and carbon to nitrogen (C:N) ratio of the plant material. The C:N ratio of plant tissue reflects the kind and age of the plants from which it was derived.

C:N ratios above 25:1 can result in nitrogen being "tied up" by soil microbes in the breakdown of carbon-rich crop residues, thus pulling nitrogen away from crop plants. Adding some nitrogen fertilizer to aid the decomposition process may be advisable with these high carbon residues. The lower the C:N ratio, the more N will be released into the soil for immediate crop use.

The C:N ratio is more a function of the plant's N content than its carbon content. Most plant materials contain close to 40% carbon. To determine the C:N ratio of any plant material, divide 40% by its nitrogen content.

For example let's say hairy vetch contains 4.2% nitrogen: 40/4.2= a C:N ratio of 9.5. A procedure for determining the nitrogen content of cover crop biomass was previously addressed in the section on nitrogen production. Estimating the nitrogen contribution of a cover crop is very helpful when adjusting N-fertilizer rates to account for legume nitrogen.

Soil and Water Conservation

When cover crops are planted solely for soil conservation, they should provide a high percentage of ground coverage as quickly as possible. Most grassy and non-legume cover crops, like buckwheat and rye, fulfil this need well.Of the winter legumes, hairy vetch provides the least autumn ground cover because it puts on most of its above-ground growth in the spring. Consequently, it offers little ground cover during the erosion-prone fall and winter period. Sowing a mix of leguminous and grassy-type cover crops will increase the ground coverage, as well as provide some nitrogen to the following crop.

The soil conservation benefits provided by a cover crop extend beyond protection of bare soil during non-crop periods. The mulch that results from a chemically or mechanically killed cover crop in no-till plantings increases water infiltration and reduces water evaporation from the soil surface. Soil cover reduces soil crusting and subsequent surface water run-off during rainy periods.

Retention of soil moisture under cover crop mulches can be a significant advantage. Dr. Blevins and other researchers showed consistently higher soil-moisture levels for corn grown in a herbicide-killed, no-till bluegrass sod than for corn grown in conventionally plowed and disked plots. They concluded that the decreased evaporation and increased moisture storage under the no-till mulch allowed plots to survive a short-term drought without severe moisture stress.

3

Agriculture Environment, Rural Sociology and Food System

ENVIRONMENTAL IMPACT OF AGRICULTURE

The environmental impact of agriculture varies based on the wide variety of agricultural practices employed around the world. Climate change and agriculture are interrelated processes, both of which take place on a global scale. Global warming is projected to have significant impacts on conditions affecting agriculture, including temperature, precipitation and glacial run-off.

These conditions determine the carrying capacity of the biosphere to produce enough food for the human population and domesticated animals. Rising carbon dioxide levels would also have effects, both detrimental and beneficial, on crop yields.

The overall effect of climate change on agriculture will depend on the balance of these effects. Assessment of the effects of global climate changes on agriculture might help to properly anticipate and adapt farming to maximize agricultural production. At the same time, agriculture has been shown to produce significant effects on climate change, primarily through the production and release of greenhouse gases such as carbon dioxide, methane, and nitrous oxide, but also by altering the Earth's land cover, which can change its ability to absorb or reflect heat and light, thus contributing to radiative forcing.

Land use change such as deforestation and desertification, together with use of fossil fuels, are the major anthropogenic sources of carbon dioxide; agriculture itself is the major contributor to increasing methane and nitrous oxide concentrations in earth's atmosphere.

CONCEPTUALIZATION OF TECHNOLOGICAL CHANGE IN AGRICULTURE

Technological change is among a small handful of topics in what is now referred to as the sociology of agriculture — along with part-time farming and analysis of the farm family — that have received attention during all three major phases in rural sociology. From the 1930s through the 1950s there were several major studies by sociologists of the impacts of technological change, most of which centered on the rapid mechanization of southern agriculture that began during the Great Depression.

The concern of many of the early rural sociologists with the socioeconomic consequences of technological change, however, largely lapsed during the succeeding phase of rural sociology in which the diffusion-adoption perspective was dominant.

In the era of the "new sociology of agriculture," we again see a rekindling of concern with the social consequences of technological change. Moreover, interest in the consequences of technological change has become expressed in more social structural terms and, as demonstrated later, has become more closely integrated with analyses of the genesis of new agricultural technologies.

Several other important observations can be made on the rural sociological literature on technological change in agriculture over the past decade.

First, mechanization has received the lion's share of attention; technological change in the forms of improved varieties of crops and breeds of livestock or of agricultural chemicals has been the focus of only a few major ex post studies.

Second, as Berardi has observed, there are many problems in decomposing the effects of technological change on farm structure from the effects of other variables such as public policy or changing wage rates; therefore, rural sociologists have often relied heavily on the primary research of others (*e.g.*, agricultural economists and geographers with complementary expertise in assessing techno logical impacts) when developing comprehensive assessments of the socioeconomic impacts of technological change.

Third, the conceptualization of technological change in agriculture by rural sociologists has been heavily influenced by the work of the agricultural economist, Willard Cochrane, who developed a theory of the "treadmill of technology" in the late 1950s. Briefly, Cochrane's theory of technological change builds on a key observation from rural sociological research on the diffusion-adoption of innovations-that large farmers who are entrepreneurial and nonrisk-averse tend to be the first to adopt new agricultural technologies.

By adopting early, these farmers enjoy "innovators' rents" because of their ability to reduce their average per unit production costs. Cochrane posits that after these innovators and early adopters have employed the new agricultural technology, aggregate output of the particular commodity involved will begin to increase and prices will decline disproportionately (because agricultural commodities tend to have a low price elasticity of demand).

Declining prices then begin to force nonadopters to utilize the technology. These subsequent adopters gain very little from the technology; they adopt it merely to be able to stay in business. Slow adopters or nonadopters will tend to be forced out of agriculture because they can no longer compete due to their high average costs of production. Cochrane considers the technological change process to be a "treadmill" in that the majority of farmers, who receive little or no benefit from the technology, are nonetheless forced to continue to adopt new agricultural technologies in order to stay in business.

Cochrane has also observed that the benefits of technological change tend over the long term to be realized in the form of rising land values, particularly if there are commodity programmes in effect to cushion a commodity sector from declining crop prices caused by increases in output.

The benefits of technological change (innovators' rents from early adoption) tend to be reinvested in farm land, causing increases in land prices and benefiting large landowners over smaller property owners and tenants. These effects, in Cochrane's view, are heightened when commodity programmes support crop prices, thereby increasing the level of innovators' rents and intensifying the extent to which these benefits are capitalized in asset values.

Cochrane's theory of the treadmill of technology has been widely accepted in recent rural sociological work as an orienting perspective because it has effectively linked knowledge from diverse origins (rural sociological research on diffusion-adoption, agricultural economics research on technological change, and so on). Cochrane's theory, however, has its limits — in particular, its inapplicability to industrial agriculture, as noted by LeVeen.

Moreover, the notion of the treadmill of technology gives little attention to the origins of technology, that is, to why technologies that have displaced labour or benefited larger operators over smaller ones have been developed, a topic that, has recently come to be a major research issue for those working in the sociology of agricultural science. Nonetheless, with the exception of sociologists of agriculture who have focused on industrial agriculture, rural sociologists have yet to develop a distinctly sociological theory of technological change and its socioeconomic consequences.

While rural sociologists have not yet developed a comprehensive and distinctly sociological theory of technological change and its consequences applicable to both family farming and industrial agriculture, the new sociology of agriculture as applied to technological change has made some significant advances. Several of the most important of these advances are summarized below.

Among the most significant research papers on the topic of technological change in agriculture were two published in the late 1970s that dealt with limitations of the classical diffusion-adoption model. The first, by Pampel and van Es, was an otherwise traditional adoption study that included, in addition to "commercial-profitable" innovations that were the major focus of previous diffusion-adoption studies, a set of "environmental innovations" such as contour farming, terracing, sod waterways, and use of rotation cover and reduced tillage.

Pampel and van Es discovered that the correlates of adoption of environmental innovations were far different than those of commercial innovations. For example, while large-scale farmers were observed to be the earliest adopters of commercial technologies, there was no significant relationship between farm size and the adoption of environmental innovations.

They concluded their study by cautioning rural sociologists that the diffusion-adoption tradition, because it had ignored dimensions of innovative behaviour such as the adoption of environmental-conservation technologies, "may have provided within rural sociology a field of knowledge with a narrow empirical foundation on which to base its generalizations". They also emphasized that given the tendency of most environmental innovations to be unprofitable, it was not clear that voluntary strategies to achieve soil conservation and related environmental goals would be successful.

Writing two years after the Pampel and van Es study, Goss focused more systematically on the limitations of diffusion-adoption research. The principal failings of diffusion-adoption research, in Goss' view, were (1) that it was uncritical, if not promotional, towards patterns of technological change that may be detrimental to the interests of society and/or farmers, and (2) that the focusing of research effort on the adoption component of the technology change process had caused rural sociologists to ignore the consequences of technological change for various groups in society.

Goss summarized the early 1970s literature critical of the Green Revolution and pointed out that the Green Revolution had been initiated with the same promotional orientation towards technology that had predominated during the heyday of the diffusion adoption tradition in U.S. rural sociology. He noted that many social scientists who had once worked within the diffusion-adoption tradition in international agricultural activities had begun to recognize the limits of the adoption paradigm. He argued, however, that this point had not been well recognized in domestic rural sociology and that rural sociology must begin to develop a more comprehensive programme of research on the consequences of technological change in agriculture.

In a manner similar to Goss' exegesis of the Green Revolution literature in terms of the impacts of technological change on social groups and classes in agriculture, Stockdale, in an influential article, raised comparable issues in terms of the ecological aspects of modern agricultural technology. Noting that the contribution of technological change to American agricultural productivity had often been held up as a model for other nations to emulate, Stockdale argued that social scientists had tended to ignore the adverse ecological impacts of energy-intensive technology in the United States and elsewhere. Like Goss, Stockdale called for an expanded programme of research on technological change in agriculture that would emphasize the social and ecological consequences of new agricultural technologies.

Among the most innovative thrusts in sociological research on technological change in agriculture have been the series of books, monographs, and articles published by Friedland and associates. Friedland and his colleagues have focused on technological change in industrial agriculture in California and have identified a number of components of the technological change process that have tended to be ignored in the earlier rural sociological literature. As noted earlier, Friedland has focused on several commodity systems (chiefly tomatoes, lettuce, and grapes) and has found that the social organization of each commodity system — and hence the structural context of technological change — is distinctive.

Nonetheless, Friedland and his associates have demonstrated that the technological change process in industrial agriculture generally is heavily influenced by considerations of the cost of, access to, and control over labour and by the relationships of grower organizations with public and private components of the agricultural research system. They have also demonstrated that the nature of the technological change process and its consequences depend upon the structure of the commodity system. For example, in the aftermath of the termination of the bracero programme in 1964, a mechanical tomato harvester technology was widely adopted in California, displacing three-quarters of California tomato producers and the bulk of their agricultural wage labour forces. In lettuce, however, grower-shippers were able to utilize the threat of lettuce harvest mechanization to discipline the labour force and to avoid the costly process of harvest mechanization.

A further component of the recent rural sociological literature on technological change, one to which Friedland and Friedland and associates have made major contributions, is social impact assessment of new agricultural technologies. Social impact assessments of new agricultural technologies are a distinctive type of research in that they often tend to rely primarily on secondary data or on the results from the published literature rather than on primary research and data collection; this is particularly the case if the assessment must be prepared over a short period of time.

Other distinctive aspects of social impact assessment of new agricultural technologies are, first, that the assessment is ex ante (rather than ex post, as in most previous rural sociological research on technological change) and, second, that the orientation towards these emerging technologies tends to be questioning and skeptical, rather than accepting and promotional. Rural sociologists who have conducted social impact assessments of new technologies have approached their subject matter from the vantage point that most new technologies tend to involve an unequal distribution of costs and benefits and that advance projections of these patterns of the distribution of costs and benefits can enable social scientists and affected groups to redress these distributional impacts more effectively.

An example of the type of knowledge base that has been developed, in part, as a building block for social-impact assessment research is Berardi's comprehensive review of the impacts of agricultural mechanization. Berardi reviewed the re sults of over 180 studies of agricultural mechanization in the United States and elsewhere and summarized the major empirical regularities in studies of a variety of types of mechanization in a diversity of commodity systems. She concluded by pointing out that the rich literature on agricultural mechanization — most of it prepared by nonsociologists — is sufficient so that sociologists may contemplate conducting rigorous ex ante studies of the impacts of new mechanical technologies in agriculture.

Unfortunately, as noted earlier, sociological knowledge on the socioeconomic impacts of mechanization is far greater than that on "biochemical" technologies (plant varieties, fertilizers, insecticides, herbicides, and so on). There have only recently been sociological studies that have delivered on Stockdale's and Berardi's admonitions in the area of biochemical technologies in plant and animal agriculture. It is likely, however, that the great attention that has already been paid by rural sociologists to emerging biotechnologies in agriculture will result in a sustained research programme into the socioeconomic consequences of these new technologies.

It should be noted in closing, however, that while there is great merit in inventorying the results of existing studies on various facets of technological change — as, for example, Berardi has done — sociologists should exercise caution, as technological change should not be assumed to have invariant consequences across time and space. In particular, as noted above, there have been some major recent changes in the structure of U.S. agriculture that one might expect will have significant effects on the technological change process and its consequences.

For example, instead of technological change tending to result in the decline of small farms, as it did during the bulk of the post-World War II period, it may be quite likely that new agricultural technologies such as biotechnology-derived plant varieties, growth regulators, and bovine growth hormone may tend to result in the disproportionate demise of medium-sized, full-time family farmers. Thus future rural sociological research on technological change in agriculture must be cognizant of ongoing changes in the structure of agriculture and how these changes affect the types of agricultural technologies that are employed, the processes of technological change, and their consequences.

THE SOCIOLOGY OF AGRICULTURAL SCIENCE

We noted several times in the preceding that the era of the new sociology of agriculture was initiated in the aftermath of a distinctly new political milieu of agricultural science and research. Hightower *Hard Tomatoes, Hard Times*, although sensational in tone and misleading on several aspects of the public agricultural research system, attracted a great deal of attention from defenders and critics of the land-grant research system. Most importantly, Hightower's book (along with contemporary ecological criticism of U.S. agricultural technology) opened up several new lines of social science enquiry into the structure and functioning of the public agricultural research system.

Probably in no other state did public agricultural science become as controversial as in California after the publication of Hightower's book. The California Agricultural Experiment Station has long had close relationships with industrial farmers and other segments of agribusiness and had been held directly responsible in the eyes of many for developing new technologies, such as thc tomato harvester, which had resulted in the rapid loss of tomato farmers and the displacement of hired farm workers. It was thus not surprising that the initial sociological enquiries into the conduct of public agricultural science were conducted in California.

The first set of studies by Fujimoto and others at the University of California, Davis, were directly inspired by Hightower and were largely focused on how agricultural scientists made research problem choices. In particular, Fujimoto and colleagues devoted attention to how research problem choices were determined by "internal" factors (such as tenure and promotion considerations) and "external" factors (such as the influence of agribusiness interests and grants). They found on the basis of personal interviews with scientists that external factors were quite important, though internal factors also exerted major influence on scientists' research priorities.

The second set of studies focused on California was initiated by Friedland and associates, who examined the processes by which agricultural commodity groups organized to exert influence on individual scientists, departments, and the University of California system as a whole in order to ensure the development and availability of technologies that would enhance their profitability. Friedland and associates gave particular attention to how the responses of University of California research officials and scientists came to emphasize mechanization and mechanization-related technology that enabled growers to displace farm workers or to enhance the control of growers over their hired labour force.

Based on the results of several years of research during the 1970s, Friedland and Kappel published a short, but still influential, monograph calling attention to biases in the public agricultural research system and outlining mechanisms by which social scientists and groups adversely affected by particular types of agricultural research could achieve changes in the land-grant system.

Shortly after the California studies were initiated, a research group at the University of Kentucky led by William B. Lacy and Lawrence Busch began a programme of enquiry into the sociology of agricultural science that continues to make major contributions to a sociological understanding of the processes by which new agricultural technologies are generated. The first publication of this group, by Busch, was a theoretical critique of the prevailing conception of scientific knowledge by researchers in the diffusion adoption tradition. Subsequent publications by the Lacy-Busch group have focused on a variety of aspects of the structure and functioning of public agricultural research activities.

Two books published by the Lacy-Busch group have had particular impact on rural sociology and the larger social science community exploring agricultural research. The first, *Science and Agricultural Development*, edited by Busch and containing several papers by his research colleagues, was the first book-length treatment of agricultural research in the tradition of the new sociology of agriculture. The chapters in the Busch anthology most germane to the sociology of agricultural science are the following: Randolph and Sachs compared the establishment of medical and agricultural science, emphasizing how a reductionist orientation and technocratic ideology had been developed in each during the process of institutionalization.

Busch and Lacy summarized the existing literature and their preliminary research on sources of influence in research problem choices in the agricultural sciences. Striving to further the social scientific conceptualization of agricultural science and technology as social products, they explored two rival social science views of the social construction and social structure of science. One, the "internalist" view deriving from Merton work, Sociology of Science, essentially holds that the outcomes of the scientific research process are determined by the "norms of science," held by practitioners and reinforced within scientific institutions, and by the very subject matter itself. The other perspective, the "externalist" view, argues that the outcomes of the research process are largely shaped by the institutions, interests, and availability of resources that originate and are determined outside of scientific disciplines and research institutions.

They suggest that both perspectives have a certain validity: "The internal and external positions are best viewed as distinct strata in the life-worlds of practicing scientists. Thus, for each discipline and institution, science is a continuous process of multilevel negotiation regarding the sources of influence on research". A final paper in the Busch collection of particular relevance to the new sociology of agriculture was Busch and Sachs', which employed a world-systems framework derived from Wallerstein to examine the establishment of the agricultural sciences in the First and Third Worlds since the seventeenth century.

Busch and Lacy are known best for their book, *Science, Agriculture, and the Politics of Research* culture, and the Politics of Research, which summarized an ambitious study of research problem choice through the use of historical methods, personal interviews, and sample survey data. Following several introductory chapters that discuss the establishment of public agricultural research, the organization of the public agricultural research system, the social demography of agricultural scientists, and the structure of communication in the agricultural sciences, they explore three major sources of influences on agricultural research: disciplinary, organizational, and extra-organizational. Busch and Lacy found scientific disciplines and disciplinary organizations to be powerful and growing influences over scientists' research problem choices.

Scientific disciplines affect scientists' research choices in a variety of ways, particularly through controlling peer-reviewed journals and establishing the norms adopted by land-grant universities and the Agricultural Research Service (ARS) in tenure and promotion decisions. Busch and Lacy also found dramatic differences among the agricultural science disciplines according to the degree to which they emphasize basic or applied research, the rate at which they accept papers in scientific journals, their norms for scientific productivity, and the degree to which scientists exhibit consensus on what are "hot" topics in the field.

Organizational context — including ARS vs. land-grant university context and the institutional prestige of the land-grant institution — was found to be a significant influence on research problem choices. Busch and Lacy found that ARS scientists were more highly constrained in their problem choices than land-grant agricultural scientists. They also found that institutional prestige was positively related to the degree to which extra-organizational influences and constraints on research choices were manifest. Among the extra-organizational influences, granting agencies and organizations (those hiring scientists on a short-term consultancy basis) were the most direct and pervasive. Whereas there were very large differences among the disciplines in the extent to which their scientists relied heavily on extramural support, Busch and Lacy emphasize that fiscal constraints on public research appropriations are increasing the pressures on scientists to obtain grants and contracts from government agencies, agribusiness firms, foundations, and so on; accordingly, scientists are becoming less autonomous in their research problem choices.

THE RURAL SOCIOLOGY OF AGRICULTURAL SCIENCE

Until the early 1980s, only small communities of rural sociologists were interested in the sociology of agricultural science. This was to change dramatically, however, in the aftermath of a workshop on biotechnology in the summer of 1982 at the University of Kentucky organized by B. Koppel, Busch, and Lacy.

Since that conference there has been growing attention to a variety of facets of biotechnology as key focal points of research in the sociology of agricultural science.

This research on biotechnology now represents a large literature, and the approaches developed therein have begun to substantially influence how rural sociologists are exploring other types of agricultural technology such as information technologies.

This new line of research has been salutary in yet another respect; sociologists interested in biotechnology have begun to make major progress in uniting the sociology of agricultural science with sociological research on technological change and social impact assessment of new agricultural technologies.

The wide-ranging and potentially integrative character of research on biotechnology, agricultural research, and agricultural change is amply illustrated by rural sociologists having given major attention to no less than 10 major themes under this rubric.

With regard to the first, the politics of the evolution of the public agricultural research system, Buttel and Kenney and Kloppenburg found that the basic science discoveries (especially recombinant DNA) relating to biotechnology, the promise of biotechnology to agribusiness firms for developing new input commodities, and federal government concern with the U.S. competitive position in the "high technologies" have rekindled scrutiny of the land-grant and ARS systems for placing too little emphasis on basic biological research.

The result was a changing political climate of public agricultural research that has led to rapid expansion of biotechnology programs in the public agricultural research institutions and to a deemphasis on traditional, applied research.

These changes in the research emphases of public agricultural research institutions have stimulated a second area of research in the sociology of biotechnology: industry-university relationships and the changing division of labour between public and private research.

Buttel *et al.*, for example, have found that industrial funding of agricultural research at the land-grant universities has increased significantly and this has generated new concerns that agribusiness firms are acquiring excessive influence on research priorities in public research institutions.

Lacy and Busch have identified a large number of likely impacts of closer industry-university relationships in agricultural biotechnology: further decline of public plant breeding (especially varietal releases), greater concentration of scientific talent at a small number of land-grant universities, decreased autonomy of public researchers in establishing their own research agendas, increased secrecy, and greater restrictions on the free flow of scientific information.

A third area of research relating to biotechnology has been on agribusiness reorganization. Kenney, Kloppenburg, and Buttel *et al.* have noted that the rise of biotechnology research has reinforced or led directly to major structural reorganization in the agribusiness industries. With the passage of the Plant Variety Protection Act of 1970, the ability of private firms to acquire patent-like protection for new crop varieties increased the perceived profit potential of seed companies, leading to the incorporation of most major independent seed companies into multinational corporations as wholly owned subsidiaries or joint-venture partners.

Biotechnology reinforced this trend, since seeds are projected to become the major commercial vehicle for plant biotechnology; moreover, the basic sciences underlying biotechnology (molecular and cell biology) are quite similar to those that will ultimately be applied in other areas of agricultural inputs such as agricultural chemicals, innoculants, and animal pharmaceuticals.

Thus biotechnology has increased the benefits of uniting chemical and plant breeding product lines within integrated multinational firms that manufacture several agricultural input products. Kenney has given particular attention to the role of small venture-capital or start-up companies in biotechnology and has analyzed the processes by which they have developed linkages with large companies.

Goodman *et al.* have stressed that biotechnology will have two distinct applications in agriculture — "appropriationism" and "substitutionism" — with different consequences for structural reorganization of the food and agricultural industries. By "appropriationism" Goodman *et al.* mean the use of biotechnology to develop new inputs, such as new crop varieties and animal breeds, for agricultural production.

Appropriationist applications of agricultural biotechnology will tend to result in the integration of agricultural input manufacturing with comparable product lines in other life science industries such as chemicals and pharmaceuticals.

"Substitutionism," on the other hand, refers to the use of biotechnology to substitute for existing products. Substitutionism will therefore have the effect of severing existing production "chains," in which specific raw materials are employed to produce particular final products (*e.g.*, substituting single-cell proteins for vegetable proteins as animal feed supplements). The implications of "substitutionist" applications of biotechnology will also include shifting food and fibre production from the farm to the factory (*e.g.*, substituting industrial tissue-cultured thaumatin for cane sugar).

Substitutionist applications will tend to lead to integration of the food and fibre processing and chemical industries.

Fourth, several researchers have examined how biotechnology has affected and will continue to affect plant breeding, including but not limited to public-sector plant breeding. Kloppenburg has conducted an extensive historical analysis of plant breeding and the seed industry in the United States. He has argued that the trends under way as a direct and indirect result of biotechnology promise to increase the penetration of agribusiness interests into land-grant university plant breeding programmes and to complete the shift, under way since the late 1930s, to curtail the role of land-grant plant breeders in developing public-domain finished plant varieties. Hansen *et al.* have reached similar conclusions on the basis of interviews with plant breeders and biotechnology researchers across the public and private sectors. Hansen *et al.* have noted as well that the increased importance of biotechnology in public-sector plant breeding programmes, combined with new avenues for agribusiness influence, will tend to result in the demise of public plant breeding efforts for minor crops.

A fifth area of enquiry in biotechnology, agriculture, and agricultural research has been on the role of intellectual property restrictions and other proprietary considerations in affecting research priorities and the nature of the new technologies that are developed.

Kloppenburg, for example, has called attention to private-sector plant breeding firms' longstanding concern about the perceived inadequacy of intellectual property institutions to protect their investments in plant breeding. In the absence of such restrictions for sexually reproduced crops prior to 1970, most seed companies focused primarily on hybrid seeds (largely hybrid corn), which, because of their reproductive instability, afforded a "natural," biological protection against farmers saving their own seeds and circumventing annual purchases of seeds from private companies.

Kloppenburg has noted, however, that the elaboration of new legal means of privatizing plant research has increased corporate interest in plant biotechnology. Kloppenburg, Kenney, and Hansen *et al.* have suggested that the role of intellectual property restrictions, combined with the fact that most agricultural input firms have both seed and chemical product lines, may lead to the development of "seed-chemical packages" that tie the farmer more directly to purchased, capital intensive inputs.

Sixth, Kloppenburg and Buttel have begun to explore the relationships between the public and private sectors in terms of recently developed "theories of the state". They have begun this task by conceptualizing public agricultural research and government commodity programmes as reflecting contradictions of state intervention.

They note public agricultural research with state intervention has tended historically to be expanded during times of socioeconomic crisis. Agricultural research, however, can be seen to set the stage for future crises in the agricultural economy by creating long-term increases in agricultural productivity. They discuss the emergence and resolution of these contradictions in the context of the Great Depression and the current era of biotechnology and farm crisis.

A seventh area of research pursued within the rubric of biotechnology has been the implications of this new trajectory of research for genetic resources. Kloppenburg and Hansen *et al.* have argued that the increased role of intellectual property restrictions combined with the new opportunities for private firms in developing new purchased input commodities will tend to lead to a decline in germplasm flows between the public and private sectors.

In the absence of greatly expanded public and quasi-public appropriations for germplasm preservation, these trends are seen to threaten the genetic resource base and perhaps to constrain plant improvement in the future. Kloppenburg and Kleinman have also examined the forces, including the rise of biotechnology, that have led to North-South conflict over control over crop germplasm.

Eighth, as noted earlier, considerable attention has been focused on how new biotechnology-related inputs such as bovine growth hormone, herbicide-tolerant crop varieties, nitrogen-fixing cereal grains, and so on might affect the structure of agriculture in the future.

Kenney *et al.* have emphasized that agricultural biotechnologies are potentially quite diverse and that how these new technologies will affect farm structures depends heavily on the research priorities that are established.

Molnar and Kinnucan, basing their work on existing theories of and evidence on technological change in agriculture, argue that larger farmers will likely be the first to adopt — and thus be the major beneficiaries of — the new biotechnologies. Kloppenburg has developed a preliminary ex ante assessment of how new biotechnologies applied to corn might affect U.S. farm structure over the coming decades.

Buttel has focused on the first major biotechnology to be applied in agriculture: bovine growth hormone, which can be produced in large quantities by genetically modified bacteria and which increases milk production per cow by up to 25 percent.

He has argued that this biotechnology will likely reinforce longstanding structural trends towards a dualistic farm structure. The Molnar and Kinnucan anthology contains several important papers on ex ante assessment of new agricultural biotechnology inputs.

A ninth area of biotechnology-related research has been in providing preliminary assessments of how biotechnology might affect the nonfarm economy of nonmetropolitan areas. Finally, Buttel *et al.*, Kenney and Buttel, and Koppel have done research into the implications of biotechnology for international agricultural research and agrarian transformation in Third World countries.

The sociology of agricultural research, especially its recent focus on biotechnology, has been conducted within the "new sociology of agriculture" tradition to a greater extent than any other area of enquiry, save perhaps for the state and agriculture.

In particular, the sociology of agricultural research, although involving some quantitative analysis, has been at the heart of trends in the new sociology of agriculture that have created legitimacy for exploring issues that are largely accessible only through historical-qualitative research methods.

Especially in biotechnology, the impacts of which on farm structure will generally not be manifest for a decade or more, quantitative research based on primary data is virtually impossible, and the quantitative research that has been done often suffers from spurious precision.

Given the importance of understanding the likely impacts of these new technologies, the work that is being done is of high priority. Some rural sociologists, however, will feel less than comfortable with a research tradition in which quantitative methodology and the putative precision it affords are largely absent.

Nonetheless, the sociology of agricultural research and biotechnology has been among the most innovative and stimulating areas of recent research in rural sociology and will no doubt have a major impact on rural sociological enquiry in the decades to come.

Agriculture and the Rural Community

Perhaps in no other area of the "new sociology of agriculture" is there so much apparent continuity with the pre-Korean War tradition of rural sociological research as in research on farm structure and rural communities.

The early rural sociologists primarily conceived of agriculture as one of the major institutional ensembles affecting rural community life and, especially during the 1930s, devoted much of their research effort relating to agriculture to understanding the implications of agricultural structural changes for rural community life and viability.

The conceptualization of agriculture and rural community interactions and the methodologies for exploring these interactions in pre-Korean War and post-1970 rural sociological research, however, exhibit dramatic differences.

Whereas most of the early rural sociologists tended to be less interested in agriculture per se and were principally concerned with understanding rural community institutions, contemporary researchers have tended to be more interested in structural change in agriculture and have deemphasized understanding the totality of the fabric of rural social life.

The research tools and data sources of these two generations of rural sociological researchers are also dramatically different. The early rural sociologists relied on rural social surveys, while the bulk of contemporary research is based on county-level census data.

A further difference between the early and contemporary rural sociologists of agriculture in their work relating to rural communities concerns the pivot of the latter studies around Goldschmidt's "As You Sow". This is somewhat ironic in that Goldschmidt's study was anthropological and qualitative, whereas the research conducted since 1970 that has drawn on Goldschmidt's work has largely been quantitative and has involved far less "anthropological" detail on rural communities than did Goldschmidt's.

Rodefeld is generally credited with the "rediscovery" and popularization of Goldschmidt's pioneering work among rural sociologists, though Heffernan (1972) made important contributions to elevating "Goldschmidt-style" research to a prominent position in rural sociology.

Nonetheless, after Rodefeld's refocusing of rural sociological attention on the Goldschmidt thesis, Goldschmidt's book has been virtually an obligatory citation in research articles on this topic. Despite some trenchant criticisms of Goldschmidt's perspective and research methodology, Goldschmidt's general hypotheses remain accepted theoretically and thus continue to shape, if not dominate, research on farm structure and rural communities within the "new sociology of agriculture."

Briefly, Goldschmidt's thesis was that there are differences in the quality of rural community life depending upon the social organization of agriculture and its attendant occupational structure.

Goldschmidt observed that the quality of rural life in two carefully matched California communities in the San Joaquin Valley differed dramatically and that these differences were largely accounted for by the organization of agriculture.

He argued that Dinuba, a community surrounded by "family" farms, had a greater number of businesses, a larger retail sales volume, a higher per capita income, and a greater diversity of social, educational, recreational, and cultural institutions than did Arvin, a community surrounded by large, industrial, corporate farms.

These differences were principally accounted for by an intervening variable — community occupational composition — that was seen to be strongly influenced by the social organization of agriculture and which directly contributed to the dramatic variations in indicators of the quality of life. Goldschmidt reported that the high ratio of laborers to other occupational groups in Arvin accounted for its low level of community viability relative to Dinuba. From these observations Goldschmidt inferred that the rise of large-scale corporate agriculture then underway in California and elsewhere would have adverse implications for rural communities.

Goldschmidt's study received essentially no further attention until it was referred to in a 1968 hearing on corporation farming before the Senate Select Committee on Small Business. LaRose and Sonka and Heady soon reported further data consistent with the Goldschmidt study.

In addition to Rodefeld's and Heffernan's efforts to resurrect Goldschmidt's work in rural sociological circles, there was a major research effort by the Small Farm Viability Project in California, drawing largely on Isao Fujimoto's unpublished research, which presented a larger restudy of the Goldschmidt thesis. The results largely supported Goldschmidt's findings of over three decades earlier. Flora and Conboy also reported results consistent with the thesis. Since that time there have been numerous studies in the new sociology of agriculture tradition that have explored the Goldschmidt thesis in a number of contexts. Several observations can be made on the conduct of recent research in the Goldschmidt tradition, what we might call "neo-Goldschmidt" research.

First, it must be stressed that contemporary rural sociological researchers have taken considerable liberties with Goldschmidt's formulations. As noted earlier, Goldschmidt's method was a comparative anthropological case study of two communities and eschewed quantitative analysis of the sort that has predominated in the subsequent rural sociological literature.

Also, Goldschmidt's analysis focused on several aspects of the property relations and control over labour, such as absentee ownership and use of ethnic minority laborers, rather than farm size, which has been the most important farm structural indicator in the neo-Goldschmidt tradition. In that regard it is important to stress that the "family" farms of Dinuba were by no means "small" farms; in fact, the family farms of Dinuba were generally larger in gross sales terms than most large farms in the Midwest today.

Second, neo- Goldschmidt research has largely focused on a single direction of causality — the impact of agricultural structure on rural communities — and has largely ignored other types of interrelations between agriculture and rural communities.

Third, most studies have been based on Census of Agriculture data and have employed countylevel data as a proxy for community characteristics have utilized survey data to infer community quality of life consequences of changing agricultural structures.

Fourth, despite obvious major regional variations in farm structures and in rural communities and their economies, there has, until very recently, been a virtual lack of a comparative perspective on farm and community structures.

Goldschmidt and Harris and Gilbert are partial exceptions to this generalization, since they have done research using states as the unit of analysis, though this procedure compounds the aggregation bias problems. Very recently, nonetheless, there has appeared a series of regionally based studies of the effects of farm structure on rural community viability and quality of life commissioned by the Office of Technology Assessment.

These studies are Buttel *et al.*, Skees and Swanson, Flora and Flora, MacCannell, and van Es *et al.* as published in abridged form in Swanson. These five studies were very briefly summarized in the main report. Together they indicate strongly that there are pronounced regional variations in how changing farm structures affect rural communities.

The impacts of farm structural changes on rural communities are generally most pronounced in the Great Plains and industrial agriculture states, less so in the South and Midwest, and very little in the Northeast. There is evidence that the degree to which farm structure affects rural communities is directly associated with the dependence of the rural community on agriculture.

The general model pursued in these neo- Goldschmidt studies has been to estimate coefficients for a causal chain in which farm structural change indicators are seen to affect the composition of the farm population (*e.g.*, as measured by indicators of the size of the farm population, the proportion of full-time hired workers) and the size and composition of the rural population.

These clusters of variables are then seen to affect indicators of the quality of life in rural-agricultural communities (*e.g.*, proportion of families in poverty, median family income, per capital retail sales, unemployment rate).

This general model is sometimes supplemented by technological factors as antecedents of farm structure, since technological change is considered to be one of the major factors influencing farm structural change.

More recently, studies in this genre have gone beyond the cross-sectional census data designs that predominated from the mid-1970s to the early 1980s and have included measures of change, where appropriate, in variables in the model.

One limit to studies that use census and other data based on single-year assessments of farm income is that net farm incomes are highly variable over the short-run; for example, as many as 20 percent of farms with annual sales of over $250,000 had poverty-level incomes in 1986.

Heffernan has effectively summarized the general configuration of results in the neo-Goldschmidt literature by noting that "all relevant research to date suggests that a corporate type of agriculture results in a reduction in the quality of community life for at least some people, especially the hired workers in rural communities."

As such, the results continue to lend support to the Goldschmidt thesis. But there have recently been several expressions of dissatisfaction with Goldschmidt's perspective and with the generalizations that have been made from research in this tradition.

A major theoretical concern is that Goldschmidt and many researchers working in this tradition have taken large-scale and corporate agriculture to be basically synonymous when, in fact, not all large-scale farms are corporate or industrial-type farms.

Much more important than industrial-type farms in farm numbers and shares of sales, assets, and profits are larger-than-family farms. Empirically, as Goldschmidt-style research has been done in the Midwest and other traditional family farming areas, there have been admonitions that large-scale agriculture, especially of a nonindustrial character, does not necessarily lead to adverse impacts on rural communities, though some studies note that increased scale of production could have detrimental impacts on community services.

RESEARCH ON FARM STRUCTURE AND RURAL COMMUNITIES

A further area of dissatisfaction with research on farm structure and rural communities is that the treatment of farm structure in this literature has generally failed to incorporate new concepts relating to agricultural dualism.

Typically farm size or scale is operationalized as a single variable representing a central tendency measure of farm size, which was a suitable procedure when there was a singular process of differentiation among farms.

But with the emerging dualism in U.S. farm structure since the late 1970s, there has been little change in mean acres per farm since the 1969 census. This measure of central tendency thus ignores the processes of dualism that have contributed to increased concentration with apparent stability in average acres per farm since the late 1960s.

The neo- Goldschmidt conceptualization of an agricultural structure in which particular communities can be characterized along a continuum of farm sizes ranging from small to large has also been questioned by Wimberley who, using multiple indicators from county-level Census of Agriculture data and a longitudinal design, found that there are instead three continua relating to degree of "corporate-commercial" agriculture, "largefarm-area" agriculture, and "small-farm" agriculture.

Using Wimberley's measures and including sources of off-farm employment as well, Reif found that large-farm-area agriculture and core-industry structures were related both to family poverty and income inequality in counties.

Thus, while studies of the impacts of changing farm structures on rural communities have been a productive area of research in the new sociology agriculture, a number of research frontiers for the future are suggested by recent criticism.

There is a need to diversify the methodological approach, consider alternative directions of causality, introduce more sophistication in terms of the measurement of change in farm structure, and expand recent efforts at comparative regional and interdisciplinary research.

The work of Vail represents an encouraging thrust in this direction. Vail has conducted extensive research on the relationships between agriculture and rural communities in Maine in which a variety of factors of demonstrated importance — the reciprocal relations between agriculture and rural communities, the role of an increasingly dualistic agrarian structure, nonfarm aspects of agriculture, and the historically specific nature of farm-community interactions — have been incorporated into a single study.

The Farm Crisis of the 1980s

The notion that American agriculture is no longer (if it ever was) isolated from the remainder of the political economy was driven home clearly by the farm crisis of the 1980s. This crisis was brought by the conjunction of a number of factors in the 1970s, including: accelerated demand for grain on world markets, public policies that encouraged massive capital investments and expansion of production, and a political-economic context that encouraged land speculation.

The farm crisis emerged in the 1980s for a number of reasons. Global agricultural export markets shrank, in large part due to the global recession. Farm commodity prices and ultimately net farm incomes (exclusive of federal commodity programme payments and price supports) declined due to the glutting of world grain markets.

Reagan administration fiscal and monetary policies led to high real interest rates, which were especially detrimental to agriculture because of its interest rate sensitivity; accordingly, debt-asset ratios among farmers increased rapidly.

The 1980s also witnessed the emergence of new Third World agricultural export competitors (*e.g.*, Argentina, Brazil) because of the global debt crisis and these countries' need to increase exports in order to repay foreign loans.

Further, federal fiscal austerity eroded the federal government's ability to maintain farm incomes through commodity programs (even though commodity programme expenditures soared during the decade). The result was a massive decapitalization of American agriculture, especially in the Midwest and Great Plains, as farmland prices in the states most affected by the crisis declined by over 50 percent.

This crisis attracted a good deal of attention by the mass media and by agricultural economists. It was not, however, until the middle of the decade that much was published about the farm crisis by rural sociologists. Most of this research can be divided into four categories: (1) neo-Marxian (and related neo-Weberian) analyses of the continuing penetration of capitalism into and the role of the state in agriculture, (2) analyses of the differences in the characteristics between farm(er)s that were and were not experiencing "financial stress," (3) studies of the effects of farm financial crisis and failure on the farm families whose enterprises were affected, and (4) studies of public reactions to the farm crisis.

The neo-Marxian (and neo-Weberian) approaches have largely focused on understanding continuing cycles of crises in agriculture under capitalism. Although these analyses focus heavily on the political and economic context of agriculture, they are not fundamentally different from what is discussed and therefore are not discussed specifically here.

Examples of this genre specifically focused on the crisis of the 1980s include Berlan, Bonanno, Mooney, McIntosh and Zey-Ferrell, Zey-Ferrell and McIntosh, Goodman and Redclift, and Buttel. We also briefly examine the farm crisis from a methodological angle — namely, the growing tendency in the political economy of agriculture to stress periods of agricultural crisis as key "test cases" for assessing theories of agrarian transition. Studies in the second category, analyses of the characteristics of farms with various levels of "farm financial stress," deal more with the effects of the political and economic environment of agriculture on farms of different characteristics. These studies have produced inconsistent results due to regional, ethnic, and other variations within the farming population.

For example, farm-size variables were found to be directly correlated with financial difficulty on Iowa farms and unrelated for Texas and North Dakota farms. Yet aggregate national data revealed that larger farms were more profitable, a relationship that obscured the underlying failure of many middle size farms. Farm families with fewer years of farming and younger ages were affected more severely.

However, Murdock *et al*. found the cohort that had started farming between the end of World War II and the mid1960s also had higher rates of financial difficulties, suggesting a life-cycle effect for those farm families bringing another generation into the farm.

The literature does, however, show agreement on the relationship between other variables and financial crisis. For example, higher debt-to-asset ratios brought by low commodity prices in a time of deflation of asset values were associated with financial problems.

The third category, analysis of the effects of the farm financial crisis and failure on the farm families whose enterprises were affected directly, has largely been the work of William and Judith Heffernan, who have conducted in-depth interviews among Missouri farm families who were in severe financial crisis or had been displaced from agriculture.

These families experienced a host of difficulties beyond the simple loss of their farms and the disorientation that accompanies the loss of a valued way of life. Among their major results are the following: Many farm families felt socially isolated and were not well-prepared for participation in nonfarm labour markets. Local employment opportunities were often unavailable, and tax laws devastated some families economically after they had attempted to leave farming with some resources with which to start a new life.

While the long-term effects of farm failure for these families were not yet known, it was clear that some families were adapting to their new situations successfully while others were at risk of suffering long-term poverty. Graham's study of displaced farm families in New York State bears out these findings. This category of studies has implications for the mutual effects of agriculture and its environment as the displaced farm families affect local labour markets and community social structures.

The final category of research on the 1980s farm crisis, on public reactions to the crisis, is best represented by Lyson's research on public apathy. Drawing on theories of public opinion formation, Lyson demonstrates that several factors led to a surprisingly high degree of apathy by nonfarm people towards the farm crisis, despite the massive media attention the crisis attracted.

He notes that there has been a trend towards farmers and consumers having very little direct contact and that the relationship between food prices and farm income has become concealed.

Most importantly, Americans have come to view "bad news" from the farm sector — the cost-price squeeze, high debt loads, and the loss of family farms — as being normal.

Thus the American public, though "interested and concerned" about the economic woes of farmers, exhibited apathy towards the farm crisis.

A major neglected area in the rural sociology literature on the farm crisis to date is studies of the effects of the farm crisis on rural communities. This topic pertains, in particular, to how conditions in agriculture affect its environment, especially at the local level.

Financial losses to local agribusiness firms when farms fail, losses of farm families in local rural market areas, declining property tax revenues due to deflated farm asset values, and other problems associated with the declining prosperity of agriculture have likely had severe effects on local services and on the local social fabric.

THE CASE OF SOIL EROSION AND CONSERVATION

Environmental problems are by no means new to American agriculture. The Dustbowl episode of the 1930s was but one of several historical instances in which concern has been raised about the environmental performance of U.S. farmers.

Nonetheless, current controversies about the management of agricultural resources in the United States date largely from the rise of the environmental movement in the 1960s and have included issues relating to soil erosion, the use of agricultural pesticides, energy and water consumption, and the like.

The soil conservation implications of American agriculture have remained among the most longstanding agricultural-environmental issues of concern to rural sociologists and are given principal emphasis.

The majority of U.S. farms are family proprietorships in which there is household or family ownership and management of the land and water resources. Accordingly, the logical point of departure in understanding the soil erosion problem in agriculture has been to conduct micro-level research on the individual, farm firm, and ecological factors associated with soil erosion rates or the utilization of soil conservation technologies.

This has been a productive line of research, but it has also become a controversial one, with the major controversies centering around theoretical approach, the relationships between micro and macro levels of analysis, and the public policy inferences that should be drawn from social science research.

Conceptual Issues in Research on Soil Erosion and Conservation

One of the most crucial aspects of soil erosion research is the very conceptualization of why soil erosion is a socioeconomic problem. Traditionally this has not been an issue, since sustained high levels of soil loss were presumed to irreversibly degrade the productivity of agricultural resources.

Accordingly, if the soil erosion problem is conceptualized as medium to long-term land productivity decline detrimental to the interests of farmers, it would follow that conservation should be achieved through voluntary farmer compliance (rather than through mandatory regulation) based on the logic of the long-term interests of farmers.

Historically, federal and state government policy thus has emphasized education of farmers and modest levels of financial inducements (such as Agricultural Stabilization and Conservation Service cost-sharing programs) to lever farmer decision making.

But it has been increasingly recognized that the proportion of prime U.S. farmland that stands to be irrevocably degraded by soil erosion is relatively small and that the "off-site" costs of soil erosion — the impacts of soil erosion and run-off on water and land resources of a farmer's parcel — are very significant and may very well be in excess of the on-site costs.

These alternative conceptualizations of the soil erosion problem — productivity loss versus destruction of off-site water and land resources — have very profound implications for appropriate types of social science research and for public policy.

To the degree that soil erosion primarily results in land productivity losses, research should focus on individual- and farm-level factors that influence resource management, and public policy should incorporate these research findings in order to develop educational and incentive programs to achieve soil conservation through farmer self-interest.

To the degree that soil erosion is primarily a problem because of off-site costs, farmers cannot be expected to conserve because the long-term productivity benefits of conservation are small, and research at a more macro level (*e.g.*, to determine the degree to which farmers' incentive structures and resource management behaviours are congruent with the public's interests in clean, navigable waters) would be most appropriate.

Debate over the appropriate kind of analysis in soil conservation research has been closely related to, and in many respects was stimulated by, debate over the adoption-diffusion approach.

Pampel and van Es as noted earlier, reported that the correlates of adoption of conservation technologies tended to be different from those of commercial nonconservation technologies. They also suggested that effective conservation practices will tend not to be profitable for farmers, so that to achieve the public interest in soil conservation voluntary compliance among farmers may be insufficient. The issues raised by Pampel and van Es have continued to pervade the erosion and conservation literature.

Many researchers have conducted their research by seeking to revise the microsociological diffusion of innovations approach, while others have argued that the constraints to soil conservation behaviour tend to be more macro in nature and that the diffusion of innovations approach will be inadequate.

Prior to 1970 soil and water conservation technologies and practices were generally understood to be a diverse collection of cropping patterns (crop rotations, contour planting, strip cropping, cover cropping, sod waterways, filter strips) and physical and biomass structures (terraces, sediment retention basins, diversion channels, animal waste structures, and hedge rows).

Since that time, however, soil and water conservation has, for the majority of researchers and policymakers, come to be largely conterminous with the adoption and use of no-till and related conservation and reduced tillage equipment, which enables farmers to leave a mulch layer of residue on the soil to reduce run-off and sediment losses.

Two observations can be made with regard to traditional and reduced-tillage equipment approaches to soil and water conservation in agriculture.

First, these two approaches, while by no means incompatible, are nonetheless quite different and will tend to apply most appropriately to different types of farm operations. Traditional cropping practices and structures are often most attractive to smaller, more diversified farmers, while reduced tillage tends to be most attractive to larger, more highly mechanized producers of row crops.

Second, reduced tillage, although acknowledged to be effective in reducing erosion in row-crop monocultures, has become somewhat controversial. One controversial aspect of the reduced-tillage approach has been that it requires increased use of herbicides, which may lead to contamination of water and thereby negate some of the run-off reduction benefits of reduced tillage.

Many studies have indicated, moreover, that reduced tillage practices have been adopted by farmers largely because of labour savings rather than because of the ability of the technology to reduce soil erosion.

Indeed, the farmers who have been most likely to adopt no-till and other reduced tillage technologies have been found to be large operators with highly specialized operations involving practices such as continuous cropping and row-crop monocultures. Thus farmer-adopters of such conservation technologies ironically represent the major trends in U.S. farm structure that have been demonstrated to lead to environmental degradation in agriculture.

However, farmer perceptions of a threat to soil productivity may lead to adoption of soil-conserving innovations, whereas stewardship related motivations may have little effect on motivations to prevent other kinds of consequences such as off-site pollution.

THE ENVIRONMENT AND THE FOOD SYSTEM

THE FOOD SYSTEM LEAVES THE FARM

Refashioning the genetic structure of crops and animals, important as it is, represents only one face of the technological revolution which has marked the appearance of the modern food system. Just as nature has been redesigned from within, through animal genetics and plant biotechnology, it has been transformed from without, by leading, in turn, to changes in the physical environment, or countryside.

These changes in the 'face of the land', themselves stimulated by technological change and accompanying shifts in the labour process have contributed to new social conflicts and political demands around what has come to be seen as the Green agenda.

The development of the food system implies a major change in the 'natural' environment, as well as in the genetic composition of the species. The scope of this change can hardly be overestimated.

The environmental effects of modern agriculture point to contradictions within the entire model on which the modern food system is based. Modern agriculture developed out of the partial separation of 'farming' activities from the natural resource base on which these activities were traditionally dependent. This led to technological innovation, and accompanying policy incentives, which have presented problems for the planned evolution of the food system.

In some sectors, such as the production of chemical fertilizers, plant breeding and pesticides, innovation developed 'autonomously', according to a scientific and industrial logic only loosely related to the cycle of agricultural production.

The reintroduction of new industrial components into agriculture, and the food system more generally, posed problems in two ways: some processes were taken out of 'farm' production (appropriation) and others were substituted, especially in food processing activities (substitution).

These transformative technological changes were associated with major shifts in the labour process, not only on the farm but in the household and in the food industry.

However, an area of equal concern has emerged during the last decade or so, which promises to pose even more serious problems for the evolution of the food system: the environment. Recent decades have witnessed a series of policy interventions designed to reduce, or reverse, the unforeseen or negative effects of the food system on the environment, largely in the industrialized countries. No area of concern demonstrates the difficulty of managing the contradictions of the food system as clearly as the environment.

Conflicts over the rural environment, and over food quality, have come to represent the clearest indication yet of the problems which the modern food system poses for sustainable development, and for the model of economic growth based on intensive accumulation which, as we have seen, spawned the modern food system itself.

This chapter examines the impact of the modern food system on the environment, in both the developed and developing countries, and the social responses which these environmental changes have evoked in recent years.

The analysis begins by considering the changes within the food system which have undermined long-term sustainability, changes which were associated with the transference of economic activities between the farm and other components of the wider system.

The major agricultural policy instruments which have been brought to bear on the rural environment in the industrialized countries are examined, notably the effects of price-support policies and economic incentives in the introduction of more capital intensive, 'high'-technology agriculture.

The effects on the environment of these policy mechanisms, including the loss of genetic resources, the pollution of water and soils, the threats to sustainable utilization posed by land degradation and the effects of agricultural change on amenity and recreation in rural areas.

These environmental effects of the modern food system are considered as first-, second- and third-order problems, each of which requires major adjustments, at the international as well as the national level.

Finally, the chapter examines the role of management interventions and environmental movements in seeking to modify or reverse these processes. A conceptual framework is outlined, which explores different forms of 'resistance' to the development of the modern food system, and concentrates attention on the epistemological and political aspects of these responses.

The description of farming in Hooten Pagnell contains most of the elements of what was still, in the 1930s, looked upon as a 'natural' farming system, heavily dependent on the cyclical and seasonal quality of rural life.

Animals were left to graze in the fields, or fed on fodder crops produced on the farm, bought inputs such as fertilizers and pesticides played a negligible part in the farm accounts, a sizeable part of the food for home consumption was still provided by the farm, marketing was still in its infancy, and the 'farm family' (largely the women) processed and prepared food in the home. Each of these elements, as we have seen, has been removed (partially or fully) from the farm economy and relocated, usually in an industrial form, elsewhere in the food system.

In contrast to this picture, modern agriculture, as we have seen, is dependent on very high levels of protein and energy transfer, linking the production process on the farm with sectors in which value is added to farm production. These linkages, both upstream and downstream, have broken into the cycle of renewal, failing to return to the resource base what has been removed from it, a principle which marks sustainable agricultural systems. Thus, in modern farming, animal waste disposal has become a 'problem' rather than an essential part of a natural recycling process.

In the same way, advanced systems of crop and animal husbandry, using every protective device from pesticides to antibiotics, are made more necessary by the conditions under which animals are kept, and the genetic make-up of delicately engineered plant varieties.

Given the commitment to high yields, consistent quality, and the major investments in research and development which underpin these technologies, once they are established there is apparently no cost-effective alternative to capital-intensive, energy-wasteful production regimes.

Nevertheless, in the process of establishing these regimes the 'knot' which tied agriculture to the environment has been un-done, and the costs of modern agriculture are counted elsewhere.

In the United States for example, it has been estimated that the cost of soil erosion off the farm was US$6.1 billion in the mid-1980s. Similarly, in the United Kingdom, £200 million in capital costs and an annual expenditure of £10 million will be required, over a twenty-year period, to bring all drinking water up to the European Community's standard for nitrate concentrations.

In the developing countries the so-called 'externalities' are often higher still. Repetto *et al* (have argued that although gross domestic product (GDP) in Indonesia grew at an annual rate of 7.1 per cent between 1970 and 1984, if this was corrected to take account of the depletion of timber, petroleum and top soil, the rate of economic growth falls to only 4 per cent, and for four of the fourteen years was actually negative.

Similarly, by calculating the effects of soil erosion on agricultural growth over ten years, Bishop and Allen have shown that in Mali soil erosion accounts for up to 16 per cent of agricultural GDP.

Costs of this sort are not included in the national accounts, and hence calculations about the benefits of agricultural growth rarely reflect the real losses sustained over long periods. There are several important respects in which attention has shifted to the problems the modern food system presents to the environment.

First, the increased dependence on differently located components of the agricultural production and food processing system has served to transfer environmental costs from one part of the system to another.

For example, as we have argued elsewhere, the use of soya meal and manioc for intensive cattle production in Europe, has transferred some of the costs of maintaining feedstock systems to other countries, such as Brazil and Thailand, which can ill afford it.

Second, increasingly intensive production has led to problems of waste disposal, pollution and disease control, all of them problems which have accompanied intensification. These problems frequently assume major importance in two directions: for the farm they represent increased costs which need to be incorporated in the farm accounts, and the market viability of the enterprise; and for the local environment they represent major threats to the existing ecological systems, species diversity and habitats.

With increasing demands for financial security, but tighter economic margins, expenditure on what are termed 'intermediate goods' now represents nearly half the value of agricultural output of some industrialized countries such as France.

Table. Intermediate Consumption in French Agriculture (Billion Francs).

	1970	1975	1981
Animal feed	6.8	14.9	33.0
Fertilizers	4.2	9.0	19.1
Fuel and repairs	3.2	6.8	15.4
Crop protection	1.2	2.9	9.6
Building maintenance	0.8	1.4	3.2
Veterinary costs	0.9	1.9	4.5
Other goods	2.5	4.8	9.5
Other services	1.4	2.4	5.0

This table, which expresses the costs of intensification both in current money terms, and as a percentage of total production costs, clearly points to the increasing financial burden of maintaining intensification, which is borne by the farmer in the first instance, but indirectly by the consumer and taxpayer and, at another remove, by agricultural competitors in less protected markets. The environmental costs of this pattern of financial and resource management on the farm are paid in a number of ways, not all of which are easily 'accountable' in monetary terms.

The third aspect of the removal of 'natural' processes from the farm, and the relocation of these processes in other sectors of the food system, has been to transform, in some cases fundamentally, the pattern of land use in rural areas. The modern food system has brought unprecedented interference with traditional land uses and the rural landscape, especially through activities like drainage, land levelling and deforestation.

There has also been a reduction in soil quality, arising from the poor management of soil resources. Current problems, as we shall see, include precipitation of acid rain on land, the discharge of toxic wastes and the uncontrolled spread of sewage on land.

As a result of these and other factors, the usable agricultural area within the European Community declined by almost 8 per cent between 1962 and 1984, from almost 100 million hectares to under 92 million hectares. The combination of several different processes, whose cumulative effect is to undermine the quantity of under-graded land available for both agricultural and non-agricultural uses (including conservation) has led to a rural resource crisis in Europe of serious dimensions. As the Food and Agricultural Organization (FAO) of the United Nations has put it, in rather august language:

A complex web of economic policies provides, at considerable cost to the environment, many of the incentives and institutional mechanisms needed to support the high level and intensity of agricultural production.

This situation is compounded by over-production of some commodities in the agricultural sector and rapidly rising economic costs from policies intended to encourage production and raise income. It is to this 'complex web of economic policies' that we need to turn next.

Policy Instruments and Mechanisms

In a document by the European Parliament itself, we are reminded of the lack of attention to long-term sustainability in the development of the European Community's agricultural policy.

Emphasizing the five fundamental objectives of the Common Agricultural Policy (CAP) of the European Community (article 39 of the Treaty of Rome)—to increase agricultural productivity, improve farm living standards, stabilize markets and ensure a sxecure supply of foodstuffs at a reasonable price—the document continues by stating that 'the pursuit of these objectives clearly betokens a productivist and materialist ideology which takes no account of environmental concerns'.

Community preference follows logically from the establishment of a single agricultural market, but it is also clear that establishing this market is expensive in terms of export subsidies 'and that it does not necessarily contribute to the rational location of different types of production'.

Similarly, within the European Agricultural Guidance and Guarantee Fund (EAGGF) of the European Community, 'guidance' to farmers (which might assist with the painful process of restructuring for farm households) only amounts to one twentieth of the total expenditure: the bulk of expenditure under this fund takes the form of subsidies to agricultural producers.

The scale of agricultural subsidies, as we shall see, makes any serious attempt at environmental protection in European agriculture extremely difficult, since farmers who are normally subsidized for doing what is harmful to the environment also expect to be subsidized for not doing what is environmentally harmful. The cost of the CAP is, as Simpson remarks 'legendary'.

Only part of the total cost of the CAP is attributable to the purchase, storage and export subsidies for farm production: a 'mere' £18 billion per annum, equivalent to £5.91 a week for a family of four in the Community. In addition, the same family has to pay almost twice this amount weekly towards the cost of fixing EC prices above prevailing world market levels. The total cost is £42 billion per annum.

These figures provide an indication of 'a policy which sacrifices long-term interests to immediate concerns, and one scarcely needs reminding that most environmental issues arise from a concern for the long-term future'. It is worth paying attention to the policy framework for agriculture within the European Community, because in intention, although not always in its choice of mechanisms, it mirrors the approach of other industrialized countries, such as the United States and Japan.

The policy framework for Europcan agriculture, as we have seen, was established with a view to protecting rural incomes through production subsidies, and ensuring food sufficiency within member countries. One consequence of these policies is that the benefits and costs of agricultural policy are shared very unequally between producers and consumers.

For example, if we compare France with the United Kingdom, we can appreciate that because of the larger size of the French agricultural sector, the absolute value of protection is higher than in the United Kingdom.

However, if we take individual farmers the reverse is true: British farmers do rather better under the CAP because they work, on average, larger farms. In France in 1980 the average farm size was 25.4 hectares compared with 68.7 hectares in the United Kingdom. As Cheshire makes clear, 'large farmers, because they produce more, effectively receive more support than poorer, smaller farmers'. The aggregate effect of these distributional factors is highly unequal: three-quarters of EC support to farmers goes to the richest quarter of farmers.

The level of expenditure under the CAP far outstripped most other budgetary items within the control of the Community. Taxpayers and consumers effectively supported 40 per cent of the value of farmers' produce during the years 1983-7, when agricultural spending in real terms increased by 4 per cent a year. In the United Kingdom, in 1984, more was spent on farmers and the food industry (£6.9 billion) than was spent on unemployment benefit and supplementary benefit combined (£5 billion).

Senior also points to the fact that only a dozen people work for the European Commission on consumer affairs, while 800 work in DG.VI, the directorate responsible for agriculture. And within the European Community it is DG.VI that makes agricultural policy, in consultation with the agricultural producers' group COPA, the Council of Agricultural Ministers and the Agricultural Committee of the European Parliament.

The effect of close collaboration between farming interests, and the virtual exclusion of the consumers' interest from food policy discussions, is that policy not only favours large farmers over smaller ones, it also favours wealthier families over poorer ones. In 1988 in the United Kingdom the poorest 25 per cent of households spent 29 per cent of their disposable income on food. The richest 25 per cent of households spent 14 per cent on food. The relatively 'rich' West Germans spent 14 per cent on food, while the relatively 'poor' Greeks spent 36 per cent.

Simpson observes that 'if poor people were paying to help other poor people and keep them on the land that may be some consolation. But they are not'. Very little influence is wielded within European Community circles by the consumer interest, and environmental considerations still do not enter into the design of agricultural policy, as one might expect.

As we have seen, this lack of political influence is in marked contrast with the position of the agricultural lobby. In explaining the ideological dominance of this lobby, we must take account of the level of farm support in the past. The level of farm support has been fundamental to the shift in agricultural practices, most of which, as we shall see, are prejudicial to the environment.

It is not difficult to demonstrate that the mechanisms, and the level, of farm support in the European Community is highly inegalitarian in its effects. Can it be demonstrated that these policy mechanisms are ultimately responsible for the environmental problems that afflict the countries of Western Europe?

The cumulative effects of high levels of farm support on the rural environment may be complex and subtle, but they are also increasingly well documented. The agricultural technology/policy model encouraged the rapid adoption of new technology in the United States and Western Europe.

The high level of farm support, expressed through high farm prices, has encouraged more investment in research and development, and the acquisition of more capital equipment by farmers. Price support has also stimulated intensification through its effect on land values.

Throughout the period between the mid-1950s and the late 1970s agricultural land values increased in the United Kingdom, in fact they tripled in real terms.

After 1979, when the rate of increase of CAP price began to fall, and particularly after 1984 when milk quotas were introduced, land prices began to decline, for the first time in decades. The intensification of agriculture was given a considerable stimulus by the capitalization of land as an asset, and the need to maximize returns on this asset.

It is also clear, from evidence such as that of Traill that high support prices for farmers induce more capital investment, and the substitution of capital equipment, where possible, for labour.

In West Germany, in 1989, there were 8.5 milking machines for every 100 cows; 2.8 combine harvesters and 12.4 tractors for every 100 hectares of land.

This level of mechanization was higher than in the United Kingdom because the farm structure was different: under a regime of high support prices, and accompanying subsidized credit and guaranteed markets, there is every inducement for even the smallest farmers to mechanize production.

Second, although the legal and financial inducements to farm amalgamation are few within the European Community, the economic gains from larger units, from 'shedding' labour and acquiring ever more capital equipment, are considerable.

The pressures on the production margins of small farmers, even those with milking machines and tractors, are further accentuated by the concentration of farm ownership. One effect of agricultural support and the technological treadmill is, therefore, greater specialization. It is in the automatic consequence of inducing investment, and losing employment, in farming.

AGRICULTURAL INTENSIFICATION

The productivist model, which lies at the heart of the modern food system, has both direct and indirect environmental effects-environmental problems in the immediate locality, and others at one remove.

The environmental problems associated with agricultural intensification are, in the first instance, experienced as a direct effect of the implementation of the agricultural technology/policy model.

As we shall see, these first-order problems hold for both developed industrial societies and for areas of modern agriculture within developing countries.

Developed Countries

The effects of agricultural intensification can be illustrated by taking as an example the removal of hedgerows. Similar accounts could be given of the effects of drainage, or the application of chemicals or intensive livestock production.

Most factors of production are in relatively elastic supply, but land is the exception. The supply of land can normally only be increased through 'improvement' or intensification.

As agriculture feels the effect of greater capital investment, prices increase and these become capitalized in land. One effect of this process is to increase the opportunity cost of unproductive land, such as hedges and field margins.

The farmer will view the conservation of hedgerows as an opportunity foregone—he has forfeited the opportunity to increase his income by producing more food. This tendency to regard conservation activities as a cost to current income is exaggerated because, under the system of grants to defray capital costs within the European Community, farmers are induced to specialize, substituting capital for labour in many of their operations.

As Cheshire argues, 'while labourers could milk cows, tend the sheep and with a simple binder, harvest the corn, milking parlours can only milk cows, and combine harvesters can only harvest cereals'. By increasing output the farmer was induced to replace organic with chemical fertilizers, and to use chemical sprays for pest control. Both sprays and fertilizers depend on subsidized machinery.

Under specialized production hedgerows were unnecessary, since they served no purpose as livestock barriers, and animals had been brought in from the fields. The use of larger machinery also meant that fields needed to be larger, persuading farmers to recoup their increased capital costs by removing unproductive hedges and ploughing over footpaths, which interfered with the movement of their mechanical equipment.

The environmental problems that have accompanied intensification are the 'effect of a system which systematically establishes financial inducements to erode the countryside, offers no rewards to offset market failures and increases the penalties imposed...on farmers who may want to farm in a way which enhances and enriches the rural environment' (. The example of hedgerow removal is just one of many similar examples of foregone production benefits being converted into environmental losses.

Among the most important effects of intensification is the use of nitrogenous and other chemical fertilizers. Within the European Community the use of chemical fertilizers has assumed increasing importance. According to the FAO, consumption of nitrogenous fertilizers in the Netherlands is 558 kg per hectare, Switzerland 436 kg, West Germany 427 kg and the United Kingdom 356 kg. There is evidence that these levels of use leave significant residues in the soil.

A report on Paris Basin agriculture shows a nitrogen surplus of 240 kg per hectare over a seven-year period. Eutrophication of lakes, rivers and coastlines caused by excessive nutrient loading from nitrate and phosphate pollution is also widespread.

The FAO notes that recent acute deterioration of marine ecosystems in the Kattegat and the Adriatic Sea, for example, has had a significant impact on fisheries and tourism in these areas.

The intensive production of animals in feedlots has had similarly important environmental effects. It is estimated that each year British farmers dispose of 200 million tons of animal manure through the river system. The disposal of slurry is now a major environmental problem.

A survey by the Water Authorities Association in July 1986 revealed that incidents of silage effluent pollution were almost double those of two years earlier.

In 1985 Anglia Water reported that half of all fish deaths leading to prosecutions, and 60 per cent of its pollution prosecutions, were attributable to slurry disposal. The excessive use of animal slurry can lead to concentrations of up to 10 kg of nitrogen per hectare. The mineralization of the soil from increased large-scale nitrate leaching can pass into both surface and groundwater. Already, in the Netherlands, the nitrate content of shallow sources of fresh water is too high to make it usable.

The average nitrate content of aquifers in the Paris Basin has doubled from 20 to 40 mg per litre during the last twenty years, while in some areas of France, such as Beauce and Sologne, the nitrate content of groundwater exceeds 100 mg per litre.

Studies of more than 6,000 wells in North Rhine-Westphalia showed that two-thirds had nitrate contents of 50 mg per litre or above, the maximum permitted under EC Directive (80/778, 15 July 1980). A third of the wells exceeded the very lenient limit of 90 mg per litre permitted in West Germany.

The cost of reducing nitrate concentration in groundwater to within acceptable limits in West Germany has been calculated as 2000 DM per hectare.

Facts such as these have led to alarm within EC policy circles. European Commission guidelines state that: 'indefinite exemption from the principle that the "polluter pays" can hardly be given to agricultural holdings, least of all those which, in terms of the way they are structured and operated, are to all intents and purposes industrial undertakings'.

The conversion of forests, pastures and wetlands to other uses has led to environmental losses in other directions, notably the loss of wildlife habitats, hydrological changes and increased risk of erosion.

Land reclamation practices have brought about radical changes in the environment during the last decades and, although the rate of land conversion is decreasing within the European Community, it is still a contributory factor in further environmental deterioration.

In the United Kingdom, for example, a dramatic loss of valuable biotopes and wildlife habitats has occurred since 1945. Almost all unimproved natural grasslands, including herb-rich hay meadows, have disappeared, as have 40 per cent of lowland acidic heaths, 30-50 per cent of lowland woods and 30 per cent of unimproved upland grasslands, heaths and blanket bogs.

As the FAO report to the United Nations Economic Commission for Europe expressed it, 'the combined effects of modification of traditional agricultural landscapes, increased use of pesticides and intensification of agriculture have reduced wild species abundance and diversity in most agricultural areas in Europe'.

There are many interrelated reasons for loss of soil productivity which result from agricultural intensification and heavy reliance on external inputs. These effects are not confined to developed countries, like those of Western Europe. The more or less proximate effects of agricultural intensification are also felt in areas of the Third World where modern agriculture has become established, often with fewer controls on adverse environmental effects than exist in the developed countries.

These areas, which we can term 'enhancement zones', perform a critical role in food production in the South, and particularly the production of food staples such as rice, which are grown under highly intensive irrigated conditions.

Through which modern intensive agriculture has served to transform the environment of developing countries, suggest that the modern food system does not need to be fully elaborated within developing countries, from both a consumption and production standpoint, for new agricultural technologies to carry important social and environmental implications there.

In the developed countries the consequences of the modern food system have been challenged on the grounds that financially underwriting modern agriculture has proved extremely expensive: the costs of agricultural support are paid by the taxpayers and consumers within the European Community, whose interests are now beginning to be championed by political leaders as well as lobbies.

The other major objection to agricultural intensification has come from the environmental groups opposed to high-input farming, or interested in preserving the amenity and recreational aspects of the countryside.

Government support for agriculture in the developed countries has been financed from domestic budgets, and the financial costs have been passed on to the population of these countries.

In the developing countries the circumstances are rather different: new developments in technology and the implementation of agricultural development policies, have been financed through external loans and credits. For example, the International Agricultural Research Centres, which form part of the CGIAR system discussed in the last chapter, are largely supported by individual governments in the developed world, development agencies such as the World Bank and by large foundations such as Rockefeller and Ford. T

o the extent that these support costs are visible, and many of them are not, it is difficult for the people of Third World countries to mount an effective opposition to the way funds are disbursed on their behalf. Similarly, people throughout the developing world are often very aware of the environmental costs of modern agriculture in areas of intensive production and of the environmental costs in 'low resource areas' which have been increasingly marginalized from the development process.

However, as we shall see, in most developing countries it is much more difficult to establish environmental priorities in the face of the 'gains' expected from economic growth, including gains in agricultural production.

Yet the heavy cost of environmental losses should not be underestimated, as we have seen. Developing countries are effectively caught in a 'scissors movement'; the costs of conserving natural resources are borne by them, while they also bear the cost of revamping their economic policies under the exigencies of structural adjustment, and other packages emanating from international financial institutions. In many parts of the Third World, especially Asia, the Green Revolution has often been depicted as a success of 'science' and 'technology' in overcoming the obstacles to increased agricultural production.

In fact, as a number of writers have indicated the most consistent factor in the production successes attributed to the Green Revolution has been public sector investment, often reinforced by externally-secured finance, and including a series of support measures affecting prices, marketing arrangements and crop procurement, the transfer of technology and input manufacture.

The effects of the modern food system have been felt, if not directly then indirectly, in the way developing countries have sought to subsidize the adoption of increasingly intensive agricultural packages.

The 'successes', in production terms, of the Green Revolution in most areas where wheat and rice are staple crops can be attributed in large measure to the transfer of the agricultural technology/policy model first developed, in its modern form, in the United States in the 1950s and 1960s.

In the developing countries the International Agricultural Research Centres, in particular, under which research institutions in the United States developed and transferred new technologies from the laboratory to the land.

The adoption of these technologies, in the developing countries as well as in the North, had more to do with the incentive structure that surrounded both their use on farms and the research and development exercise.

One important aspect of the expansion of the modern food system is that components of this system, if rarely the system in its entirety, have served to transform the context of agriculture in so many Third World countries, despite structural constraints to the adoption of the fully elaborated industrial food system.

Essentially, the adoption of 'high' capital-intensive technology under conditions in which most of the rural population live in material poverty, has in many cases served to exacerbate inequalities within regions.

Under these circumstances, ecological problems, especially increasing fragility of the ecological system, are closely related to a reduction in livelihood choices for poor people, and their efforts to re-establish a viable livelihood are frequently costly in environmental terms, whether in resource poor areas or in areas of capital-intensive agriculture.

Recent papers from the World Bank have suggested some useful policy options, which would enlarge the management interventions available to government. One example is the removal of pesticide subsidies, which are costly in economic terms, for most developing countries, and undermine the stability of agriculture.

The removal of these subsidies, or the introduction of a pesticide tax, would reduce pesticide consumption, and give farmers an increased incentive to adopt more sustainable, biological alternatives (such as those indicated under the general rubric of Integrated Pest Management).

In the case of Indonesia, it has already been demonstrated that reducing pesticide subsidies brought fiscal savings of US$150 million annually over a period of four years. As World Bank economists have shown, the reduction of these subsidies contributed to more pest-resistant rice varieties becoming available, and Integrated Pest Management practices became more widespread.

Marginal Lands

The secondary, or indirect, effects of the food system on the environment in both developed and developing countries are more difficult to delineate. The process through which the modern food system becomes established at the farm level—concentration of holdings, specialization and intensification of production—carries implications for areas other than those of modern agriculture.

The cost of achieving gains in productivity in one locality is often passed on to others; through environmental costs or 'externalities'. As we have seen, intensification needs to be understood not simply in terms of the technology employed but also in terms of a 'package', comprising guaranteed prices, subsidized chemical and mechanical inputs and government procurement after the harvest.

In some developed countries, such as those of the European Community, less advantaged regions receive subsidies from central government precisely because they are adversely affected in social terms—they are less competitive in terms of their economic performance.

In developing countries, however, areas where natural resources are poor, especially semi-arid regions, are likely to fall completely outside the ambit of public policy. This neglect serves to accentuate the differences, the unevenness of the development process.

It is more difficult for small farmers, for example, to gain government support for their farming system; they are left to manage an inadequate technology, even if this technology is environmentally sustainable, ill-adapted for competition with the technology employed in more favoured regions.

Many households, in marginal areas of Europe, Australia and North America, such as the Appalachian Mountains, as well as the Third World, have to supplement the income they obtain from their land by other forms of livelihood.

They become part-time farmers, working in the city or in tourism, or migrating on a seasonal basis to other regions of the country. Living in areas that are already poor in terms of productive resources, but frequently rich in terms of natural species or aesthetic quality, they are powerless to do anything in the face of the differentiating effects of modern intensive agriculture.

This chapter is concerned with both the immediate and the more indirect environmental implications of the modern food system. It needs to be emphasized that a series of 'third-order problems' also exist which we are only beginning to delineate: that is, problems such as those arising from the introduction of genetically engineered species into the environment.

The discussion of biotechnology and genetic engineering touched on these issues, but it is too early to elaborate at length on the social implications. These 'third-order problems' are likely to grow in urgency and include not only the compatability of 'new' life forms and existing ecosystems, but also the relationship between conserving species' diversity in laboratories, and gene banks, as well as in the wild.

The Management Response

The early part of this chapter addressed the question of agricultural policy instruments, and the way these had been used to support intensification, at the cost of environmental considerations.

Given the high visibility of environmental issues and, in the developed countries, the acknowledgement that environmental considerations require changes in agricultural policy, it has become important to review the various ways in which the rural environment can be more effectively 'managed'. Beginning with an analysis of environmental management in rural Britain, and extending this to the rest of the Community, this section goes on to assess the kinds of farm-level response which have been induced by recent policy making.

The question posed by this discussion of sectoral and farm-level decision-making, is whether it carries implications for developing countries, where the penetration of the modern food system is also eliciting environmental management interventions. This means looking very carefully at the assumptions underlying environmental management itself.

As Potter observes, the increasing concern with long term support for farming in Britain and the European Community generally has 'coincided', especially in the British case, with public concern about the environment and its conservation.

As we have seen, policies which encourage agricultural intensification, specialization and the concentration of farm ownership in fewer hands have been called into question by the existence of farm surpluses and the heavy financial cost of supporting the agricultural sector. It needs to be emphasized, however, that it is public involvement in conservation, and the lobbying of environmental interest groups in Britain, which has called the model itself into question.

The agriculture and environment debate in the United Kingdom dates back to the early 1960s, although it did not reach the European stage until almost twenty years later. Beginning in the 1960s the public's attention was alerted to the cumulative effects of using pesticides in agriculture. The Nature Conservancy began to reveal that there had been dramatic losses in some wildlife species, such as peregrine falcons, due to the use of organo-chlorines. Within a decade this concern, fostered by the growth of the countryside lobby, had extended to the loss of whole habitats, such as moorland, heath-land and ancient woodland, losses which were attributed to agricultural intensification and land reclamation.

Groups such as the Council for the Protection of Rural England (CPRE) which had considerable following, especially in the more scenically attractive parts of the British countryside, drew attention to these environmental losses. Other controversies were more long-standing but no less vituperative: the ploughing of moorland in the Exmoor National Park, for example, drew attention to the limited powers exercised over supposedly 'protected areas'.

Land use controversies began to engage public attention as much as environmental pollution or contamination from intensive farming practices. Furthermore, the effect of EC membership in accelerating these losses gave rise to additional concern at the environmental costs of 'productivist' agricultural policies.

The loss of the countryside's amenity value was a particularly British concern. As Potter shows, the agenda was partly set by relatively autonomous bodies known as 'quangos' within the structure of the British government, notably the Countryside Commission and the Nature Conservancy Council.

In 1974 the Countryside Commission published its 'New Agricultural Landscapes' study, which documented the ways in which familiar features of the British countryside, such as hedges, copses and ponds, had been removed, as farmers deserted mixed husbandry for more specialized farming operations.

By 1977 the Nature Conservancy Council (NCC), largely staffed by trained environmental scientists, with no obvious 'political axe' to grind, had reported that existing policies, which sought to protect sites of special environmental interest rather than more integrated land use practices, were seriously deficient.

By 1984 the NCC had revealed that only 3 per cent of lowland grassland had remained unaffected by agricultural intensification, since 1945, and that ancient lowland woodlands had been reduced by between 30 and 50 per cent in the same period.

While the environmental lobby was beginning to raise its flag during the 1960s and 1970s, evidence was emerging that the agricultural lobby was being forced to adapt, slowly and painfully, to the realization that it could no longer count on the unqualified support of British governments.

Two agricultural White Papers in the 1970s departed, in significant ways, from the conventional wisdom that, as the major land-using industry, agriculture could expect to expand unchecked. Although no revised national policy existed which integrated the agricultural restructuring being urged on the European

Community with more attention to environmental objectives, questions such as the appropriate size of the agricultural industry in terms of food and fibre output had become matters of policy debate. As a recent research paper makes clear, a subtle shift had occurred between 1975 and 1979, in which the Ministry of Agriculture conceded that the 'national interest' was distinguishable from that of farmers.

By 1981 the tide of concern over the demise of the British countryside led a Conservative government to introduce a Wildlife and Countryside Act which provided a legal basis for management agreements between farmers and the NCC on farmed land of special scientific interest. This Act allowed for payments to be made to farmers, by way of compensation for forcgoing modern agricultural practices, in return for agreed conservation practices. Owners and landowners were required to seek permission from the NCC if they intended to change their farming practices.

The scheme attracted considerable criticism for being 'too little, too late', resting on the farmer's goodwill, and proving very costly to the Exchequer. Although the shift away from unbridled productivism had occurred, the Government seemed anxious to proceed by agreement and co-operation with farmers, almost whatever the cost. Voluntary and informal structures of land use planning and regulation were very much the method favoured by the Government.

It is this concern with 'management agreements' that marks out recent policy interventions in the industrialized countries. According to a recent report from the OECD there are three desirable aspects of such agreements.

First, the environmental benefits have to be realised effectively. For example, if the aim is to prevent cropping, then the cropping rights should be permanently acquired under the agreement.

Second, the payment made in compensation for the reduction in expected net farm income, which the adoption of new management practices is likely to produce, should not exceed this 'opportunity cost'.

Third, the agreements should not be in force long enough to enable farmers to bargain about the level of compensation, by refusing to operate under the new conditions until all their demands are met. It will be clear from these 'desirable conditions' that it would be impracticable to introduce similar agreements on a large scale in most developing countries.

The mounting controversy about agricultural support, however, did begin to alter the policy debate by linking the redirection of agricultural support to rural conservation measures.

Within the EC structures legislation and the enabling legislation contained in the 1986 Agriculture Act, financial aid began to be offered to appropriate farming practices as well as to farmers who refrained from using inappropriate practices. A number of areas considered 'environmentally sensitive' were outlined in the 1986 Act, and about 750,000 hectares of land were designated in the United Kingdom.

Farmers were being paid, for the first time, to 'produce' countryside, something which, they did without significant agricultural support in the era before agricultural intensification.

The principle that the interests of farmers were distinguishable from those of the British countryside was conceded in the 1986 Agriculture Act, which 'imposed a statutory duty on Agriculture Ministers to balance the conservation and promotion of the enjoyment of the countryside, the support of a stable and efficient agricultural industry and the economic and social interest of rural areas'.

Although the agricultural lobby no longer commands undivided support, even from Conservative governments in Britain, and the environment lobby has increased in size and authority, the mechanisms through which agricultural support can be shifted towards better conservation are still in their infancy.

Writing about the United States, where the conflict is more often between agricultural intensification and sustainable agriculture, rather than'countryside' with amenity and recreational value, Buttel argues that farmers are most unlikely to change their practices to meet environmental guidelines: State policy-makers have been unable to call upon farmers to collectively and voluntarily reduce their production during phases of overproduction.

Given the inability of farmers to control their production, state action to reduce production by rationing land resources has been undertaken because rationing other inputs (ie. petrochemicals) would be strongly resisted by farmers and by agribusiness firms.

Some commentators, such as Potter have argued that the preservation of existing farms is a necessary component of any wider conservation goals in European rural areas, and that the offering of income aids to farmers who agree to make their farming practices more environmentally sensitive is the most direct and powerful way of advancing future rural environmental policy. However, the same writer is equally aware of the limitations of such a small-farm approach.

He writes that: realistically, though, there are limits to the extent to which a Farm Survival Policy can be orientated towards conservation objectives. With the notable exception of environmentally sensitive areas it is unlikely that the disadvantaged and intermediate farms that will be targeted in future will be located in areas of greatest environmental potential or vulnerability.

The very transparency of expenditure on income support presents another difficulty: it may not be desirable to seek to achieve long-term environmental goals under expenditure headings that may be vulnerable to political pressure.

It is not clear, then, that the situation in Europe is as strikingly different from that of North America as first appears. The powerful economic lobbies behind high-input farming have not been seriously affected by the new calls to conservation. Indeed one feature of so-called 'Green consumerism' today, is that commercial interests with dubious environmental records are often loudest in their support for more 'sustainable' practices. The political climate has changed in recent years, and this is not an unimportant fact in itself, but the model which encouraged agricultural intensification as part of the development of the modern food system, has been dented rather than overturned. In the process, 'management' has acquired both greater credibility and a more uncertain future.

MOVEMENTS AND SUSTAINABLE DEVELOPMENT

Most of the recent discussion of 'environmental management' in developing countries has paid scant attention to the role of local people. Environmental management is defined almost exclusively terms of a rigidly 'top-down' approach. It is assumed that by incorporating better systems of environmental accounting in the national accounts, and by ensuring that the negative costs of development are allowed for within project budgets, progress will be made towards better management of the environment.

Desirable as it is that economics takes environmental problems seriously, such measures are unlikely to affect the balance of forces which cause poor rural people to place excessive strain on their environments. The 'official' discussion of environmental management in the South makes barely a reference to the increasingly important experiences, and literature, on environmental movements there. The remaining section of this chapter focuses attention upon the need to re-examine our knowledge and understanding of the environment, in the light of the experiences of other cultures far removed, socially and geographically, from the 'core' of the modern food system.

Sustainability and Development

It is clear that rural development projects have often neglected local, ecological and cultural factors in the South. It is also clear that structural factors—which influence people in not acting 'sustainably'—have frequently been ignored.

We need, in fact, to take account of both the cultural and structural dimensions in any definition of sustainability. Then we can 'begin to ask whether there are patterns of myth and social organisation, which are more likely to ensure and maintain man's *[sic]* survival base'. It is clear that environmental problems are not new. Historical, and archaeological, evidence suggests that early societies often behaved in ways which were damaging to their environment.

These societies also had to 'solve the basic problem of producing food surpluses and collecting raw materials from rural areas to sustain large urban populations', a problem that faces countries in the South today. However, such societies, unlike the South today, were not faced by twin assaults from rapid economic and demographic growth, on such a scale. They did not exist within a global context which placed their own food systems and values in jeopardy.

If we are to understand fully the relationship between sustainability and development, we must begin by recognizing that they are paradigms with historical as well as geographical antecedents which need to be located within specific social and economic contexts.

The point at issue is that societies differ in the way they prescribe courses of action towards the environment. Too often we assume systems possess universalistic characteristics, including the food system.

Such universalistic approaches to social and political systems often leave no room for human agency within the framework of explanation. The question we might address in exploring environmental movements was posed by Giddens in the following way: how does the 'practical consciousness' that people acquire from their society, and from their environment, relate to the 'structural properties of the system'? Environmental movements include movements of resistance to the encompassing food system. Under what circumstances are the structural properties of such a system called into question?

The Epistemological Dimension

To answer this question we might begin, not with an analysis of political conflict over the environment, but with the way in which the 'environment' is understood.

Instead of working 'backwards' from social movements to consciousness, we can work 'forwards', from the formation of environmental consciousness to the organization of alternative forms of social action to redress environmental problems.

It is clear that the way in which knowledge is gained about the environment and the way in which this knowledge is used have considerable bearing on the question of human agency, and with it environmental management practices.

Howes has argued that in agricultural systems engaged in monocrop cultivation the world view is one in which 'society is conceived as an entity apart from Nature'. It is clear, too, that many non-Western cultures possessed 'scientific theory' before Western science, as we know it, had fully developed. Looking at examples from South Asia, Goonatilake claims that, 'their large-order cognitive maps anticipated the non-manipulable, and distant reality of modern science'.

Furthermore, these kinds of knowledge system often codify knowledge in ways that ensure its cultural survival, long after the reasons for certain observances have passed.

In some cases 'after an initial map has been made and legitimized…it tends to continue to be believed for considerable periods of time'. A useful distinction, although it should be emphasized that it exists *within* societies as well as between them, is between what Feyerabend calls *historical traditions* and *abstract traditions*. The two traditions can coexist in a society at any point in time, but some cultures give more emphasis to one tradition than to the other.

Historical traditions involve a different discourse from abstract traditions. It is assumed that 'the objects already have a language of their own' and it is the task of people within a society to learn this language.

Their task is to 'learn the language of the objects as they are, and not as they appear, after they have been subjected to standardizing procedures' and experimentation. According to this tradition there is no such thing as pursuing 'objective truth', since truth involves both an object and a subject. Feyerabend compares this with what he terms an abstract tradition.

Abstract traditions are different in that they involve framing statements as a means of building up a model of what happens in nature. According to Feyerabend, 'the statements are subjected to certain rules (of logic, testing and argument) and events affect the statements only in accordance with the rules' It is possible, then, to make scientific statements without personal contact with the objects being described.

Both these epistemological traditions exist, at different levels, in most societies, but within the Western scientific tradition much more emphasis is given to abstract traditions.

In this respect the development of Western science is best understood in the context of Judaeo-Christian thinking, which has often appeared inimical to 'scientific' method.

For example, Goonatilake argues that religious and philosophical objections to scientific method in Europe would not have occurred in South Asia, where most religious beliefs emphasize the role of both abstract and historical traditions in arriving at valid statements about the world around us.

The first point that needs to be emphasized in establishing a framework for understanding environmental conflict in the South is that a society's epistemological framework will influence its understanding of the environment. What we perceive as 'environmental' problems may be interpreted in other cultures as essentially due to the breakdown of social and cultural practices and beliefs.

In addressing these questions we need to be aware that our understanding of Societies very unlike those of the industrialized countries frequently distorts our understanding of their environments.

By the same token, the rediscovery that knowledge-systems other than our own have importance for understanding our relationship to the environment has become a central, but contentious, tenet of recent Green thinking.

As Luke observes, some Green thinkers have misunderstood the importance of non-Western science. 'Deep ecologists' have sometimes equated small-scale, non-Western societies with 'primal societies', implying that 'primitives' provided an echo of our own pre-Enlightenment past.

This rather naïve view of the superiority of non-Western epistemologies fails to recognize that myths, magic and ritual are 'the functional equivalents (and perhaps the conceptual antecedents) of Enlightenment science and technology'.

This kind of simplification and misunderstanding also errs in lifting ecophilosophies out of their cultural context—Taoism or Buddhism—'with little consideration of any cultural grounding', when it is the specific cultural context of environmental epistemologies that gives them their force and vitality.

As an approach, the close association of indigenous knowledge with 'answers' to current environmental problems is frequently ethnocentric, a historical and reductionist, substituting descriptions of the world for more rigorous and less ethnocentric analysis.

Other examples exist of societies closer to our own in which environmental problems are constructed from very different political and social milieux. In Eastern Europe today, for example, the commitment to radical political ecology, and the growth of Green movements, is also linked to processes that are not strictly 'environmental' in the narrow sense employed in the West Conflicts over ethnic identity, and the struggles for civil and human rights in contemporary Eastern Europe, have incorporated—and adapted to—environmental demands, and often express these demands in language drawn from the discourse.

The Economic Dimension

The second, important dimension of relations between the environment and society is the economic dimension. The environment is subjected to structural changes over time, linked to economic growth and the development process.

If we confine ourselves to the relationship between agriculture and the environment we can identify several features of the development process which have major implications for environmental systems, among them: the production of food as a commodity rather than for household consumption; the adoption of 'high-technology' energy systems on the farm, and in the food processing industries, rather than 'low-technology' energy inputs; and the role of technology in reducing the pressure exercised by population increase on limited natural resources. These factors will be looked at in turn.

Food as an Exchange Value

In some societies agricultural practices underwrite sustainability, in the sense that consideration of the long-term implications of current agricultural practices is enshrined in farm management practices. For example, Wilken shows how some small farmers in Central America, without recourse to fossil fuels, effectively 'bank' their own labour, and that of their household, investing labour in resource management practices that make sense within a longer time-frame. They are seeking to reduce future operational costs on the farm by making major current investments in labour-intensive management now.

As we have argued above, in some societies such practices become ritually encoded. Rappaport shows how, under stable conditions, the ritual enactment of myths may not only help to maintain the social cohesion of the group (a point often made by anthropologists) but also help to sustain the natural environment. Of course, the value placed on a resource is linked to other social referents, such as the concept of ownership, as well as economic factors like the cost of extraction.

The development of wider markets for food products, and the adoption of production techniques that depend on sophisticated off-farm processing and industrial inputs, does not necessarily mark the point at which food becomes a commodity—in most societies food has been a commodity for a long time. However, it does mark the point at which the conversion of food from a use value to an exchange value has the greatest impact on the way the environment is managed, and ultimately socially constructed.

Food is taken from the province of reproduction in small-scale societies and transferred to that of production and consumption in more complex societies. Eventually, in what we can term 'post-industrial societies', the locus of dissent shifts again, this time from production relations to those of consumption. Some of the most important implications for the evolution of the food system are represented by resistance to the introduction of new, processed 'convenience' foods.

The Political Dimension

The development of the modern food system has brought new urgency to another important dimension: the political dimension. The wider parameters of the food system have introduced this dimension into social relations on the farm, relations which were previously mediated by more endogenous factors.

Linkages between agricultural producers, suppliers and markets have necessitated more joint decision making, and have led farmers and the food industry into political alliances. As producers and consumers have become incorporated into wider market systems, their response to their tacit incorporation has often taken the form of resistance, in which class, gender or ethnic subordination is contested.

The 'fields of resistance' that result have often served as rallying points for disaffected groups with different political agendas. The concept of fields of resistance we take from Foucault. It is used to distinguish between different responses to the 'individualising techniques of power', through which systems are able to exert social control and often develop a political consensus. The individual's response to the management of this consensus lies at the heart of the resistance to industrially produced food in the developed countries. Attempts to create a consensus around modern agriculture and food policy have recently broken down.Among the forms of resistance which environmental movements express we can include:

- *Exploitation,* through which the individual is controlled via market forces and the production process;
- *Domination,* through which the individual's identity is defined in terms of gender or ethnic roles; and
- *Subjection,* where the controlling effect of dominant belief systems in the society tie the individual to an ideological position.

An analytical framework based on fields of resistance is of use in exploring the political changes that accompany the development of the modern food system, and the transformation of natural resources.

For example, in the case of farmers in the United Kingdom the growth of market forces and production relations on the farm proved to be the decisive factor, at the turn of this century, in bringing about national organizations of both farmers and farm workers from different class positions. However today, farmers and landowners are united in their resistance to restructuring brought about by the market, as well as to government policies on conservation or 'countryside management'.

Farmers have become dependent on the state, and at the same time resentful of their dependence—an ambiguity behind the attraction of agrarian ideology, especially in countries like France and Germany, where farms were smaller and the dislike of state corporatism more pervasive.

By contrast, the importance of *subjection,* as a field of resistance, has grown with the increasing effect of modern agricultural practices on the environment. The demand for more 'organic' food, the interest in preserving the physical condition of the countryside, and political mobilization around issues like animal rights and food quality have brought a new urgency to the interface between agriculture and the environment.

The political dimension exists in any set of environment-society relations. However, we need to ask why resistance sometimes finds expression through collective activity (land-rights, urban struggles, etc.) and why it sometimes implies withdrawing from the wider public policy sphere into an 'alternative' community or individual 'survivalist' position, at odds with the mainstream political system.

These questions inevitably raise wider issues concerning the conflicting uses to which the environment can be put. The conflicts that surround resistance to 'development' bring into the open alternative strategies for utilizing local knowledge as well as conserving nature. The analysis, at this point, cannot be confined to the food system, and embraces the trajectory under which autochthonous alternatives to development diverge from mainstream economic forces under the direction of capital located in the North. These wider issues are discussed at length in Redclift.

Having considered the principal features of the modern food system from a historical and cross-disciplinary perspective, it is necessary to take the analysis forward into the realm of contemporary culture and politics.

This involves analysing the growth of opposition to changes in the food we consume. This chapter has been principally concerned with exploring the effect of the developing food system on rural land use. Increasingly, the focus of concern, for interest groups, politicians and the consumer, has shifted to what we eat, and changes in our diet in turn reflect wider changes in consumption and the use of leisure.

The food system assumes its most radical and yet most familiar form in the supermarkets where we in the developed countries buy most of our food. Resistance to changes in food marketing, processing and retailing have assumed a greater importance in recent years. Today they represent a veritable counter revolution.

Food and Energy Transfers

The food system also affords an excellent example of the way that energy use is linked to the capacity of ecosystems to retain, as well as transfer, value. The process of development, as we have seen, is usually considered in relation to the movement of capital and labour, rather than the exploitation of natural resources.

The structural and spatial aspects of transfers of value from one part of the (developing) world to another part of the developed world have been emphasized by O'Connor and Smith. Without dissenting from the significance of these forms of surplus extraction, it is important to emphasize that transfers of energy from within one system to another can also be seen as transfers of value.

Bunker uses both ecological principles, and the laws of thermodynamics, to assert that 'our calculus of value must include not just the labour and capital incorporated into commodities, but all forms of energy and matter'. In his view, 'the embodiment of energy in economic and social organization encompasses far more of the essential differences, and relations, between core and periphery, than measures linked to commodity production and exchange can'.

Population Increase and Carrying Capacity

Economics has also experienced considerable difficulty recently in incorporating the full implications of changes in environmental systems within its analytical framework. The Malthusian tradition emphasized that population pressure placed limits on the capacity of natural resources to feed the population. In fact, like many economists, Malthus did not have a great deal to say about the environment.

Barbier reminds us that the pessimism of the Malthusian tradition was only matched by the optimism of the Ricardian view of scarcity. This placed emphasis on 'relative scarcity' rather than 'absolute scarcity'; as resources became used up substitutes were found for them, extending the time-frame in which the environment could facilitate economic growth.

The problem of carrying capacity was not solved but it was 'put back', and technology was the major factor in effecting this change. Later neo-classical thinking, represented by Alfred Marshall, took an even more optimistic position than Ricardo; no tendency was predicted for profits, wages and rent to decline with increased scarcity of natural resources.

The implication was that we were entering an 'age of substitution', the environment could be exploited without degradation being inevitable, indeed the scarcity of non-renewable resources might attract a higher price, and therefore lead to their better conservation.

This view of 'substitution' still plays a large part in the way we conceptualize the environment today, and has been given added impetus by developments in biotechnology.

The importance of the economic dimension is that it necessarily involves considering the conversion of natural capital into social capital, economic processes of growth and development do not simply modify, they actually transform the environment.

In the course of these transformations the development process 'hands back' to nature some serious problems, including that of sustaining economic levels of production and consumption.

ENVIRONMENTAL MANAGEMENT

André Gorz makes a distinction in his work between what he calls 'environmentalism' and 'ecologism'. Environmentalism refers to planned intervention to secure improvements in environmental quality, much of it undertaken by the state in developed countries.

Ecologism represents much more deep-seated objections to the way the environment is handled in such societies: 'Ecologism is not content with correcting the environmental impact of the system. It examines the reasons *why* certain technologies and forms of production and consumption were promoted in the first place'.

In the developed countries, as we have seen, the state already intervenes to correct or modify the full effects of the development process on the environment. In such countries, the 'core countries' for the modern food system, the popular mass diet is already well established, and agriculture employs only a small proportion of the population. Under these circumstances, managing the wider environmental implications of the food system assumes importance largely because basic needs have been met.

Environmental policy interventions usually take two forms as far as food system activities are concerned: the protection of 'valued' environments from the pressures of human economic activity and the reduction of market pressures on specified groups of agricultural producers, to enable them to act more 'sustainably'.

Activity during 1984	Percentage of people (aged 12 or over) taking part in each activity.
Drives, outings, picnics	55
Visits to historic buildings	39
Long walks	33
Visits to country parks	27
'Pick your own' (PYO) farms	23
Nature reserves	13

Note:

- Countryside Commission, National Countryside Recreation Survey, *SocialTrends* 17.
- In any four week period during 1984, 40 per cent of the population made no visits to the countryside, while 17 per cent made nine visits or more.

Table. Membership of Environmental Organizations.

Organization	1970	1975	1980	1985	1990
National Wildlife Federation	2,600	n.a.	4,600	4,500	5,800
Sierra Club	114	153	182	363	566
National Audubon Society	105	255	310	425	515
Wilderness Society	66	n.a.	63	97	363
Environmental Defence Fund	10	40	45	50	150
Greenpeace	"	6	80	450	2,000
National Resources Defence Council	"	15	35	65	140

Both kinds of intervention are, arguably, holding devices, since they can be continuously negotiated and renegotiated. In fact the more interventionist the state becomes the larger and more vocal the environmental lobby usually becomes, exposing new environmental risks and dangers, to which it is claimed government should be responsive.

The principal objectives of environmental policy, in relation to agriculture, speak to the priorities and concerns of urban middle-class groups: amenity recreation, species preservation and aesthetics. The membership of environmentalist organizations has increased rapidly in recent years. In developed countries the existence of protected areas does not diminish opposition to environmental management decisions, but it does draw attention to some of the problems and limitations of what we have described as the 'modern food system' model.

In developing countries the situation is usually markedly different. Environmental conflicts in developing countries are about poor peoples' access to the resource base in the struggle to maintain livelihoods, and the conflicts which are associated with changes in resource use in the course of economic development.

'Sustainability' in this context is not primarily a consumption goal of the relatively leisured middle class, but a concept which, if it is espoused by anyone, is espoused by the rural poor themselves and usually within the context of human rights. Of course, this is not always the case, and the fact that the rural poor often stand in an ambivalent position to more sustainable systems of agricultural production is evidence of the difficulty in securing livelihood advances.

It is worth emphasizing, too, that the interest of people in developing countries in their own environments does not put a seal on environmental conflict in these societies. There are also issues of the 'international commons': common property resources, as well as open access resources to which no property rights apply. Many of the world's life-support functions are dependent on the protection of resources in large tracts of the South; genetic materials provide the best example, as we have seen, but increasingly even issues like rain forests, climatic change and the disposal of toxic wastes involve relations between the developed and developing countries.

The globalization of the modern food system ensures that changes in resource use, and the environmental effects of new technologies, are transmitted from one geographical location to another, just as patterns of food consumption have become increasingly international.

Policy and management interventions to secure more 'sustainable systems' cannot be judged simply in terms of their own logic and internal dynamic; they need to be considered within the broader context of production, consumption and the transformation of resources which defines the modern international food system.

As the British response to the World Conservation Strategy made clear, 'any useful analysis of the reasons for natural reserve destruction would have to include questions of wealth distribution *within* and *between* nations'.

The point about environmental management, as it is practised in the developed and developing countries alike, is that it takes as a 'given' the distributive consequences which the market produces in the course of development. In developed countries, as we have seen, increasingly vocal and well-organized groups have mobilized around environmental objectives. In developing countries only the poorest and most marginal are likely to do so: rubber tappers in the Brazilian Amazon, Chipko activists in India, and women farmers in the Kenyan 'Green Belt' movement. The politics of environmental movements, as well as management interventions, have a different basis in the South:

The more aid flows towards those sectors in developing countries favoured by political, industrial and commercial criteria, the less likely they are to arrest such environmental damage as is occuring.

This is because the greatest environmental damage, and needs, tend to occur in the rural areas and amongst the urban poor of the least developed countries (with the obvious exception of industrial pollution which is not insignificant in some developing countries). In the developed countries the management of the rural environment is contested by two opposing interests: those of farmers and conservation bodies.

The environmental lobby has assumed increasing importance in the United Kingdom because environmental values are located in both resource-rich and resource-poor areas. Although some relatively remote areas are under populated and offer few attractions to agriculture, most rural areas are within reach of heavily populated urban settlements and land uses are strongly contested.

In North America, Australia and parts of Europe the environment presents itself in another form, that of wilderness, which has served as a focus for the environmental lobby, both as intellectual inspiration ('deep ecology') and as a barometer of cultural, and hence political, concern. Nash, in *Wilderness and the American Mind,* called these areas of wilderness 'institutional containers', vital because they have 'resisted' civilization.

The juxtaposition of dense urban populations and resources worthy of conservation, in the developed countries, enhanced the importance of 'protected area' status, such as that of National Parks. Most of these areas had little agricultural potential, but enormous cultural and political weight was attached to their conservation. The difficulties afflicting areas of some agricultural importance, which contained or bordered zones of environmental value, were rather different. Strategies have had to be evolved for reducing productivist pressures in these areas, either through 'setting aside' land deemed surplus to agricultural requirements or by shifting towards 'lower-input' farming.

Beginning with physical planning controls, and protection of areas designated as being of special interest, the move within the countries of the European Community has been towards more economic incentives to practice 'sustainable agriculture', including measures to support farm families as well as farmed landscapes. The motivation to undertake more management interventions has been provided not only by heightened environmental concern, but the onerous budgetary costs of supporting agriculture and production of food surpluses. The comparison with the developing countries makes it clear that environmental management principles cannot simply be transferred from North to South. In developing countries the contest over natural resources, their use and ownership, is between the rich and the poor in rural areas.

In most cases environmental agencies are influenced by external pressures—such as those associated with opposition to tropical deforestation—and, in any case, such agencies have no 'teeth'. Protected areas such as Biosphere Reserves are demarcated, but the means with which to ensure their protection are rarely available; budgets are minimal, 'policing' activities random or unenforceable.

In areas other than protected areas there is little, if any, outside intervention to protect the environment. In some cases, agricultural technology is disseminated which is more 'environment-friendly', such as that of Farming Systems Research (FSR), but this technology is encouraged largely because high-input agriculture is impossible on marginal or resource-poor land.

Only relatively recently have 'sustainable' objectives begun to be incorporated in production programmes for small farmers, but the vast majority of the rural population, particularly rural wage labourers, lie outside the reach of existing 'sustainable' policy packages.

It is extremely unlikely that policies like the 'polluter pays' principle could be incorporated within the gamut of policies influencing rural resource uses in developing countries. The pressures mounted within the international community to call a halt to environmental degradation in the South have had little impact on the economic policies and practices that largely govern North-South relations.

One much-debated exception is that of 'debt-for-nature' swaps, in which some areas of tropical forests are 'protected' from development, and their value written-off against part of the country's external debt.

In practice, however, such swaps feature in only tiny areas of endangered land, it is unlikely that they can be adequately protected, and in any case the decision to do so was not arrived at democratically but by administrative fiat.

We have yet to see even the beginning of effective local participation, by indigeous communites and others, in the long-term sustainable management of tropical forest areas. What has emerged from the 'debt-for-nature' debate is a measure of hostility between some governments in the South, anxious to protect their sovereignty and forced to increase foreign exchange earnings by selling products that carry environmental costs, and what are sometimes perceived as 'interfering' First World environmentalists.

Developing country governments argue, not unreasonably, that they cannot be expected to practise sustainable development when their paymasters, the developed countries, fail to do so.

4

Agriculture and Crop Plants

AREAS OF PLANT DIVERSITY

Biologists, ethnologists and geneticists pay tribute to the genius of Nicolay Ivanovich Vavilov - his enormous width of understanding and innovative vision. We may well consider him to be the 'Darwin of the 20th century' with just as enquiring a mind and capacity to recognize the basic similarities between several apparently quite distinct phenomena as Darwin had done in the 19th century. Whereas Darwin studied the diversity of all living organisms known to science, Vavilov directed his vision towards domesticated plants just as had Alphonse de Candolle before him.

Clearly, Vavilov's depth of enquiry was much greater than that of de Candolle, to whom he pays tribute. Vavilov was basically a geneticist and plant breeder, approaching the problems of cultivated plant species in terms of the diversity within and between them that might be put to practical ends. For this reason Vavilov was not particularly interested in diversity as such, but only in the diversity that could be put to practical advantage.

The world flora, even a quick survey that there are many areas of plant diversity which have little to do with cultivated plant origins. Thus, the 'fynbos' plant formation on the very southernmost tip of South Africa is extremely diverse. In quite a small area one can find hundreds of species in quite distinct genera and even plant families.

These are very attractive and colourful but were never brought into cultivation by the indigenous people. They were not at all edible and are known today as beautiful and extraordinary examples of plant evolution.

A rather similar flora exists in southern Australia, and again, none of its species was brought into cultivation. the vast tropical rainforests of South America, Africa and Asia.

These differ from the fynbos in that many plants have been used, perhaps for millennia, for food, medicine, clothing and in some instances building materials. But were any of these domesticated? Perhaps a few, such as cassava, pineapple, peanuts, etc., but these were plants from the drier forest margins, and occurred only in certain regions. Several nuts and fruits were gathered for food. They were gathered, eaten and used in a variety of ways. Again, the vastly rich fruit-tree floras of South East Asia were of immense value to the people of such areas but their actual cultivation is comparatively recent.

The wild spring bulb flora of the central and southern European mountains. These plant communities are very diverse and extremely beautiful, with their bulbs, annuals, shrubs and trees.

Few were eaten by humans - in fact the plants themselves had developed mechanisms to prevent their being eaten, such as poisonous or bitter roots, bulbs, and spiny leaves and branches.

None of these was domesticated in ancient times. Much further north in northern Europe and the steppes of Asia, as well as the northern part of North America and the southern part of South America, the species diversity diminishes very greatly. There are only a few species able to survive, grazed by wild and domesticated cattle, but these are not of direct use to humans as food, only through their horses and cattle.

There is considerable wild plant diversity up to about 45-50°N latitude in Europe, Asia and North America, and southward to 35-40°S latitude in the southern continents, apart from desert and semi-desert areas. However, only in certain areas between 45° and 30°S latitudes were potential crop plants actually domesticated.

The centers of origin of cultivated plants occurred mostly imountainous regions between the Tropic of Capricorn (23°28') south of the equator and about 45°N of the equator in the Old World. In the New World crop domestication occurred between the two tropics (Cancer and Capricorn) approximately. In all cases agricultural origins and primitive diversity occurred in high and complex mountain regions. Why only these?

The Process of Domestication

Domesticated despite their evolutionary processes to protect them from being eaten by the development of various defensivemechanisms such as poisons, bitter substancesOne strategy of survival adopted by many plants is to produce so many seeds that even if most of them are eaten enough will remain to provide for the next generation. And the plants that do this superbly well are grasses.

To a lesser extent those that do this quite well are herbaceous legumes. This, then, may be the solution to the problem of how humans began the process of domestication. Hunter-gatherers took and ate the natural surplus and left enough seeds (probably by chance) to provide for the next generation. Later, when non-shattering mutants occurred by chance, these were adopted automatically by the primitive farmers. From this point evolution under domestication began to take place.Vavilov considered that "as a rule the primary foci of crop origins were in mountainous regions, characterized by the presence of dominant alleles

CENTERS OF ORIGIN AND DIVERSITY

In this he recognizes eight primary centers, as follows:

The Chinese Centre: In which he recognizes 138 distinct species of which probably the earlier and most important were cereals, buckwheats and legumes.

The Indian Centre (including the entire subcontinent): Based originally on rice, millets and legumes, with a total of 117 species.

The Indo-Malayan Centre (including Indonesia, Philippines, etc.): With root crops (*Dioscorea* spp., *Tacca*, etc.) preponderant, also with fruit crops, sugarcanc, spices, etc., some 55 species.

The Inner Asiatic Centre (Tadjikistan, Uzbekistan, etc.): With wheats, rye and many herbaceous legumes, as well as seed-sown root crops and fruits, some 42 species.

Asia Minor (including Transcaucasia, Iran and Turkmenistan): With more wheats, rye, oats, seed and forage legumes, fruits, etc., some 83 species.

The Mediterranean Centre Of : More limited importance than the others to the east, but including wheats, barleys, forage plants, vegetables and fruits - especially also spices and ethereal oil plants, some 84 species.

The Abyssinian (now Ethiopian) Centre: Of lesser importance, mostly a refuge of crops from other regions, especially wheats and barleys, local grains, spices, etc., some 38 species.

The South Mexican and Central American Centre: Important for maize, *Phaseolus* and Cucurbitaceous species, with spices, fruits and fibre plants, some 49 species.

South America Andes region (Bolivia, Peru, Ecuador): Important for potatoes, other root crops, grain crops of the Andes, vegetables, spices and fruits, as well as drugs (cocaine, quinine, tobacco, etc.), some 45 species.

The Chilean Centre - Only four species: Outside the main area of crop domestication, and one of these (*Solarium tuberosum*) derived from the Andean centre in any case. This could hardly be compared with the eight main centers.

Brazilian-Paraguayan Centre: Again outside the main centers with only 13 species, though *Manihot* (cassava) and *Arachis* (peanut) are of considerable importance; others such as pineapple, *Hevea* rubber, *Theobroma cacao* were probably domesticated much later. the very wide range of plant diversity in the tropical and warm temperate regions of the world our major food crops have come mainly from high mountain valleys, isolated from each other to a large extent and with a very great habitat range.

Here people made selections of wheat, barley, oats, rye, potatoes and maize which were eventually cultivated. These plants were weeds or possessed the syndrome of not being able to compete well with climax vegetation. Hence they grew in areas where nature or humans had reduced competition from other species, were noticed, eaten, resown by chance and eventually became domesticated. Several other weedy plants were never or only temporarily domesticated, remaining as weeds but often hybridizing by chance with the cultivated ones and thus enhancing their diversity.

The restricted access of the mountain valleys and the wide range of altitudes helped to produce and select the diversity needed for domestication. Similar selection pressures even in unrelated crops produced similar types of adaptation, a process developed by Vavilov into his Law of Homologous Series.

Because such adaptation in only partially related crops must surely have been due to mutations on distinct loci in each crop, this writer feels that a more correct title might have been the Law of Analogous Series.

However, the phrase has persisted as Homologous Series and we must retain it as just one of the extraordinarily innovative ideas put forward by the great genius, Weedy relatives of the crops were also present, adding to the genetic diversity by random mating with the primitive crops themselves. In the crop I know best -potatoes - one can find not only a range of ploidy, from diploid to pentaploid, but clear evidence for crosses between weed species and cultivated ones as well as the natural introduction of frost resistance from wild species into the cultivated ones.

That evolution of domesticates is very much promoted by the factors of varied microclimate, aspect, altitude, restricted habitats and human selection which are all present in the intertropical mountain zones of the Old and New Worlds. Such a wide range of natural and human selection pressures is not available to the same degree of intensity in other world regions.

Vavilov was the first investigator to study and understand these systems and to use them as a basis for present and future plant breeding. At the same time, since Vavilov was not only a geneticist and plant breeder, but also a man of wide interests and intelligence, he was able to provide a theoretical background to interpret crop genetic diversity in all its aspects and thus to make it available for the whole world.

THE CENTERS OF TYPE-FORMATION (CENTERS OF ORIGIN) OF CULTIVATED PLANTS

The centers of type-formation or the centers of diversity the 'differential phyto-geographical method:

- A strict differentiation of the plants Linnaean species and intraspecific groups by all available means of various disciplines beginning with morphology, agrobotany, phytopathology, cytology and recently by molecular methods.
- Delimitation of the distribution areas of these plants and, if possible, also of the distribution areas in the remote past when communication and seed exchange were more difficult than at present.

- A detailed determination of the composition of the varieties and races of each species, and a general system of the genetic variability within the different species.
- Establishment of the distribution of the genetic variability of the forms of a given species as far as regions and areas are concerned, and the establishment of the geographical centers where these varieties are now accumulated. Regions of maximum diversity, usually also including a number of endemic types and characteristics, can also be centers of type-formation.
- For a more exact definition of the centre of origin and type-formation it is necessary to establish the geographical centers of concentrations of species that are botanically closely related as well.
- Finally, the establishment of the areas of diversity of wild subspecies and species that are closely related to the cultivated species in question should be used for amendment and addition to the area defined as area of origin, when the differential method for studying races is applied to them.

In 1940 Vavilov stated that the method of differential taxonomy offers an opportunity to trace the dispersal of many cultivated plants. It demonstrates their stages of evolution with respect to both the initial origin and their introduction into cultivation within different areas.

It shows their relation to wild subspecies and species, but also demonstrates the subsequent evolution under domestication of these plants when dispersed from the basic centers and undergoing changes under new conditions and the further effects of natural and artificial selection.

The origin of different cultivated plants led Vavilov to the establishment of new concepts, *i.e.* primary and more ancient crops in contrast to secondary ones, allowing him to characterize with good precision the centers where agriculture originated and the pathways along which it was dispersed.

The geographical distribution of plant resources on earth and the establishment of the enormous infraspecific diversity of the majority of crops allowed not only a determination of their localization but also offered an opportunity to ascertain the period of origin of the plants most important for cultivation. "The history and origin of human civilizations and agriculture are, no doubt, much older than what any ancient documentation in the form of objects, inscriptions and bas-reliefs reveals to us.

A more intimate knowledge of cultivated plants and their differentiation into geographical groups helps us attribute their origin to very remote epochs, where 5000 to 10,000 years represent but a short moment" The number of centers listed in the terms 'center', 'focus' and also 'area' of origin. Their definition is important: the geographical centers are basic and independent foci where agricultural crops originated but are also geographic areas where cultivated plants are grown the centers of origin of cultivated plants but they amount only to a correction of details.

"The basic composition of cultivated plants, typical of this or that centre, remains stable" As far as the historical character of Vavilov's works towards the establishment of the centers of agricultural crops is concerned, Sinskaya calls our attention to the prevalent use of the expressions "historical-geographical area" or "geographical areas of the historic development of a cultivated flora" Sinskaya elaborated a more detailed approach to the analysis of the cultivated plants in their centers of origin.

This approach is based on a differentiated characterization of the endemism the various taxa have in a given area which are divided into the following categories:

- Genera originating from the area.
- Genera having one of their centers of origin or their most important secondary centre of origin in the given area.
- Species strictly endemic for the given area.
- Species endemic to the given area, but having their first origin in another area.
- Species having one of their centres of origin or their most important secondary centre of origin in the given area.

The main categories and proposed to differentiate five basic geographical areas of historical development of cultivated plants, each having its subareas. The basic geographical areas of origin of cultivated plants after the respective cultivated plants together with their characterization by the above-named categories.

The development of important genera (such as *Triticum, Secale, Hordeum, Beta, Brassica, Daucus, Lens, Linum, Olea, Mandragora, Pisum, Melilotus* and many others), which include the major proportion of all crops, forms the basis for agricultural production in countries around the Mediterranean, but most of them are intensively grown in Asia as well as in Southwest Asia.

Phytogeographical studies have revealed that compared with the areas of Africa south of the Mediterranean, there is a characteristic cultivated flora that is not less rich than those in other centers where agriculture arose. Many cultivated plants have undergone very old but secondary development there, *e.g.* in Ethiopia.

While further developing Vavilov's ideas about the centers of origin, singled out the African region for the historic development of cultivated flora. Ancient Mediterranean elements (actually both Mediterranean and Southwest Asiatic ones) predominated in the composition of the cultivated flora of Ethiopia but are not sharply delimited from those of other African areas.

Elements from South Asia also occur there. This area is rather an area of introduction than of distribution of cultivated plants to other places. calls such territories "dependent areas", to which belong not only Ethiopia, but also North America, where agriculture developed on the basis of Mexican and Central American crops and, later on, that of crops from the Old World.

In central and northern Europe, on the Russian steppes and in Siberia, agriculture is based primarily also on cultivated plants, introduced from the subareas Southwest Asia and Mediterranean, etc.

The agriculture of the "dependent areas" underwent a certain period of development and, therefore, is not limited, judging by the large quantity of plants introduced into cultivation from less rich, wild flora of these territories.

Table. Geographical Areas of Historical Development of Cultivated Flora.

	Basic areas of origin	Subareas
I.	Ancient Mediterranean	Southwest Asia1a. Anterior Asia (Transcaucasus, Asia Minor, Near East, West Iran)1b. Middle Asia (Turkistan, Afghanistan, East Iran, North West India, Pakistan)Mediterranean.
II.	East Asia	Northeast Asia (Japan, Manchuria) Southeast and Central China.
III.	South Asia	South China, India and Sri LankaMalesian.
IV.	Africa	
V.	New World	Central AmericaSouth America.

Vavilov referred emphatically to the division of the globe into floristic regions and subregions such as those accepted by conventional phytogeography for elaborating the geographic origin of cultivated plants. The origin of cultivated wheat is located in the Ancient Mediterranean (syn.=Old Mediterranean) which includes, the Mediterranean region and Southwest Asia. The latter was divided by Vavilov into Asia Minor in a broader sense and Inner Asia.

Sinskaya considers these territories as subareas of the Ancient Mediterranean Many species, genera and families which were responsible for the development of cultivated plants originated from these subareas. Nevertheless, every subarea has its characteristics due to the ecological conditions and the richness of the flora of wild and cultivated plants as well as to the ancient history of agriculture.

The area of origin of wheat is Southwest Asia. The greatest amount of endemic species and a huge amount of different intraspecific taxa is found there. The greater the distance from this primary area of origin, the less diversity of the species is observed.

From 26 species of the genus *Triticum* the following species are found in Anterior Asia and are endemic wild plants: *T. urartu, T. araraticum* and *T. dicoccoides.* Endemic cultivated plants are: *T. timopheevii, T. zhukovskyi, T. carthlicum, T. karamyschevii, T. ispahanicum, T. macha, T. vavilovii* and *T.*

sinskajae. Triticum turanicum and the wild einkorn *T. boeoticum* mainly occur there. Sinskaya considers *T. sphaerococcum* a further species, which occurs in northwest India, as an endemic species.

A second group of Southwest Asian wheat originated in the Near East subarea but then spread to other areas. They were later replaced, at the beginning of the century, by higher-yielding species and therefore can only be found as relics isolated from each other. They are: *Triticum monococcum, T. dicoccum, T. aethiopicum* and *T. spelta* (Sinskaya 1969; Padulosi *et al.* 1996). These species, which are at present to be found in areas far away from each other, were more intensively developed in other subareas of the Ancient Mediterranean. *Triticum aestivum* and *T. compactum* were intensively developed in Middle Asia and subsequently spread all over the world.

Triticum durum and *T. turgidum* developed in the more central and western parts of the Ancient Mediterranean, particularly close to the sea coast. East of this area of origin, on the other hand, the process of formation of the further wheat varieties is not observed. A.M. Gorskyi during his expedition to Sinkiang (Western China) found a new endemic wheat named *T. petropavlovskyi* (Dorofeev *et al.* 1979). Vavilov had completed the investigation of Sinkiang in 1929 and regarded this western part of China as one of the geographically

PLANT PRODUCTION

Over the past ten years much new evidence has to come to light which has enabled us to explain in more detail the transition from plant-gathering to plant production in Southwest Asia.

It is now clear that this important change, which led ultimately to a significant increase in population, urbanism in Mesopotamia and Egypt, the civilizations of Greece and Rome and eventually industrialization, occurred gradually over a long period in a geographically wide area. These plant remains themselves have provided several hundred radiocarbon dates.

Archaeobotanical evidence indicates that the process of domestication may have been slow and finds indicate that domestic and wild cereals occurred as mixtures on several early period of at least a millennium Archaeobotanical finds and field studies show clearly that late Epipalaeolithic and early Neolithic distributions of wild cereals were much more extensive and that the cereals collected which favoured certain cereals (barley on poor dry soils, for example).

These differences can be seen prior to and just after cultivation began. But once cultivation became systematic, favourable soils would have been chosen as would preferred crops. For example, emmer became more widespread at the expense of einkorn.

METHODS

In southwestern Asia from the crucial period 20,000 to 8500 BP (non-calibrated). The quantity and quality of the archaeobotanical Because biological decomposition is rapid in the aerobic archaeological sediments of this area, archaeobotanists rely on plant materials which have been rendered stable through charring in hearths or other fires. These remains are recovered by flotation and sieving.

Under the best circumstances large-scale flotation has obtained thousands of charred seeds, fruits and fragments of charcoal within the chronological At worst no sampling was carried out or only a few chance finds were collected. Concerning the archaeobotanical criteria for morphological domestication and cultivation, not all archaeobotanists agree on the best criteria.

The solid rachides in barley and naked wheats are clear indicators for archaeobotanists. However, the more primitive hulled wheats such as einkorn and emmer are more problematical. For the archaeobotanist the distinction between domestic and wild hulled wheats is based on the disarticulation scar left by the abscission layer.

The break occurs in the same place on the rachis, and on domestic modern material it is rough or torn and on wild material it is smooth. However, with ancient carbonized remains the surface is very often too poorly preserved to allow this distinction.

Grain size is another criterion used, because domestic grains are generally more plump Plump barley grains, unknown in the wild, occur with fragile rachis fragments These are not considered to be evidence of domestication. However there is some reason to reconsider these finds in the light of modern semi-solid rachis barley which occurs in Syria.

The author has collected specimens of semi-solid rachis, two-row 'black' barley near Bosra in southern Syria. The disarticulation scar is similar to that of wild barley, and the rachis fragments would be difficult to distinguish from wild types in carbonized material.

This morphological type could explain why domestic-type grains occur with apparent wild-type rachis fragments. As for evidence for cultivation, one might expect digging tools to provide the answer. However, for the moment this is not the case and it is possible that these tools were wooden and have not survived. Archaeobotanical research has concentrated on the presence of weed assemblages

The most common taxa in these assemblages include the following: *Adonis, Aegilops, Astragalus, Avena, Bromus, Bupleurum, Camelina, Centaurea, Centranthus, Coronilla, Fumaria, Galium, Glaucium, Hordeum, Lathyrus, Lithospernum, Lolium, Malva, Papaver, Polygonum, Reseda, Silene, Valerianella* and *Vicia.* It is difficult to be sure that these taxa really represent a weed assemblage because these plants make up part of the original steppe flora, and identification at the species level is rarely possible.

On the other hand, what one might expect is an increase in the frequency of these taxa at the expense of other steppe plants which were not preadapted to become part of the weed flora. Food grains have the immense advantage that they can be stored. The wild grasses become ripe in late spring and promptly fall to the ground. They need to be harvested just before maturity. Given the dry climatic conditions, grains can be conserved or stored relatively easily.

This facility for storage was one of the main advantages of seed-gathering which led to a more secure subsistence base and prepared the way for a sedentary way of life. Increasingly favourable climatic conditions resulted in a rich environment for Early Natufian inhabitants of the western Mediterranean (12,000-11,000 BP).

The climate appears to have been more favourable than at present or at any time since the Natufian. Archaeobotanical evidence indicates that the Syrian and Jordanian steppes had a much richer vegetation. This is indicated by the presence of forest steppe species, for example *Pistacia* and *Amygdalus.*

Cultural factors such as an increased reliance on stored grain (although there is little archaeological evidence for this) permitted a sedentary existence, such as Mureybit and Abu Hureyra on the Euphrates in Syria, Hayonim in Israel, Wadi Hammeh in Jordan, and a little later Qermez Dere, Nemrik 9 and M'lefaat in northern Iraq have round architecture, large hearths and groundstone equipment.

That wild cereals were exploited together with a number of edible fruits and pulses Charcoal and fruit remains indicate the forest/steppe with *Pistacia* and almond; in favourable conditions, deciduous oak was present. Archaeobotanical evidence clearly indicates that this vegetation penetrated further east into what is now arid steppe.

This habitat provided wild cereals, pulses and an abundance of game the use of grasses comes from glossed flint tools indicating the harvesting of plants with high silica content Hatoula and Kebara in Israel and Beidha in Jordan It is clear that morphologically wild progenitors of Old World cereals and legumes were exploited for several millennia, before the appearance of their domestic counterparts.

The geographical extent is impressive, stretching from northern Iraq to the southern Levant, Anatolia and even southeast Europe. During this period Einkorn is dominant at Mureybit and Abu Hureyra, barley and some emmer are present at Ohalo II.

Rye is also present at a number of the Late Natufian there is wide evidence for climatic deterioration from about 11,000, usually referred to as the Younger Dryas (which appears to have adversely affected settlements in the more arid zones of the Jordanian and Syrian steppe and the Negev highlands.

With few exceptions, near permanent water continued to be occupied into the next period. During the following period, the Pre-Pottery Neolithic A, the climate became more favourable again. Charcoal evidence indicates that the

Syrian steppe was at least partly wooded near reliable water sources. The architecture of small round houses is more substantialFor the very earliest levels at Aswad IA and Jericho, remains are numerically too meagre to be certain of domestication, Emmer is dominant at Aswad einkorn at Mureybit, barley at Jerf al Ahmar in northern Syria suggests that the inhabitants of these were still gathering local cereals but this does not exclude small-scale cultivation as described by Harris using locally available wild cereals as seed stock.. The controversial plump domestic-type barley grains

Established during this period or earlier show no clear-cut domestication in the lowest levels, but morphological domestication does appear at higher levels. over a very long period, more than a millennium. This would have led to degradation of the local vegetation Thus resource depletion could have been a contributing factor for the adoption of agriculture. During the next chronological period architecture becomes rectangular and the transition. near. At Aswad, between 9730 and 8560 BP, 26% of the barley rachis fragments are solid domestic types, but it is not clear whether they occur in the earliest levels. At preliminary studies indicate that the cereals are not yet domesticated, but indirect evidence of weed associations strongly suggests the presence of cultivation, and similar assemblages are seen at Aswad, Cayönü and Cafer Höyük.

More frequent and cover a wider geographical area with expansion of agricultural communities into central Anatolia (Asikli Höyük) and Cyprus (Shillourokambos).

Crop evolution and morphological domestication are clearly shown by the appearance of a solid rachis in barley and naked wheat, for example at Aswad West phase II In the Euphrates Valley there is no evidence for *in situ* domestication unless we consider the plump barley Instead domesticates were introduced.

Emmer absent in the area during earlier periods appears to have been introduced from elsewhere at Abu Hureyra and Halula together with a naked wheat wild types remain at significant frequencies. We can see this at Cayönü, Cafer Höyük (wild wheats), Aswad, Ganj Dareh (wild barley), Halula and Azraq (wild wheats and barley) These mixed finds can be interpreted in three ways: (1) as evidence of the exploitation of wild stands, (2) as unwanted weeds, and (3) as an integral part of the crop consisting of a mixture of wild and domestic cereals.

The relatively high proportion of wild types and the lack of pure finds of domesticates suggests that the wild plants may have been considered as a useful part of the crop, as opposed to unwanted weeds. This suggests cultivation of wild and domestic types together but does not exclude gathering from wild stands in a kind of mixed economy. Even during later periods for example at Ramad between 8210 and 7880 BP, domestic barley rachis fragments are only at 52%. A similar situation was noted at Magzalia such as Bouqras and Ras Shamra (phase Vc) wild types are rare or absent. wheat. During the Late PPNB, einkorn becomes a minor component and could be interpreted as a weed for most of the Near East.

DOMESTICATION OF PLANTS

Lentil (*Lens culinaris* Medikus) is a short-statured, annual, self-pollinating, food legume mainly grown in the Indian subcontinent, the Mediterranean region and North America. The crop is grown in dryland cereal-based rotations because of its nitrogen-fixing ability, its high-protein seeds for human consumption, and its straw, which is a valued livestock feed. ICARDA has a global mandate for research on lentil improvement and is situated in the Near East arc, where the crop was domesticated. The temporal dimension given by the history of the spread of the lentil following domestication and the consequent selective forces give a perspective of applied evolution to plant breeding which is highly relevant to ICARDA's current role in any new 'spread' of the crop.

LENTIL ORIGINS

The putative progenitor of the cultivated lentil is *Lens culinaris* subsp. *orientalis* (Boiss.) Ponert which is distributed from Greece in the west to Uzbekistan in the east, and from the Crimean Peninsula in the north to Jordan in the south The oldest carbonized remains of lentil are from Franchthi cave in Greece dated to 11,000 BC and from Tell Mureybit in Syria dated 8500-7500 BC But as it is not possible to differentiate wild from cultivated small-seeded lentil, the state of domestication of these and other carbonized remains in the aceramic farming villages in the 7th millennium BC in the Near East arc is unknown.

The finding of a large hoard of lentil at Yiftah-el dated to 6800 BC is, however, suggestive of domestication The oldest find of lentil seeds that are larger than wild seeds, and therefore unequivocally domesticated was at Tepe Sabz, Iran; they have been dated to 5500-5000 BC. The overlap in the distribution of wild lentil and the early archaeological record indicates that lentil was domesticated in the Near East arc.

Spread

From these beginnings the crop spread to the Nile, and to Central Europe via the Danube. As lentils are repeatedly found in the early agricultural settlements of the 5th millennium BC in Europe, situated outside the distribution of *L. culinaris* subsp. *orientalis,* this is indicative of earlier domestication. Lentils were definitely associated with the start of the 'agricultural revolution' in the Old World, which was initiated by the domestication of einkorn and emmer wheats, barley, pea, flax and lentil The crop was part of the assemblage of Near Eastern grain crops introduced to Ethiopia by the invaders of the Hamites. From the Bronze Age onward, maintained itself as an important companion of wheat and barley throughout the expanding realm of Mediterranean-type agriculture.

The dissemination eastward of the Near Eastern grain crops, including lentil, reached Georgia in the 5th and early 4th millennia BC. The crop appears in the archaeological record in India around 2500 BC as part of the Harappan crop assemblage.

Alphonse de Candolle wrote that on linguistic grounds, "It may be supposed that the lentil was not in this country (India) before the invasion of the Sanskrit-speaking race." The invasion occurred before 2000 BC. The crop probably reached its current Old World range about 3000 years ago. It was carried to the New World after Columbus.

Selective Forces Operating During Spread

the selective forces operating during the spread of the culture of lentil derives from an analysis of the variation found in current landrace populations from different geographic regions.

Following widespread collecting and evaluation classified the assembled variation into six groups (grex varietatum), each of which was geographically differentiated and also characterized by a complex of morphological characters, mainly qualitative, common within a group but differing in other groups.

This type of geographic association is mirrored by variation in quantitative morphological traits Such morphological characters are readily observable and consequently often the subject of human selection. Geographic differentiation between landraces also has been found for other more cryptic, ecophysiological factors, such as those resulting from selection for soil conditions and for climatic conditions, as the following examples illustrate.

Iron-deficiency symptoms are observable on some accessions of lentil grown in calcareous soil. In a germplasm collection of 3512 accessions originating from 18 countries, landraces from those Mediterranean countries where lentil originated (Syria and Turkey) exhibited Fe-deficiency symptoms only at very low frequencies ($<1.5\%$) In these regions the soils are generally highly calcareous with pH >8.0, conditions known to reduce Fe availability.

Those landraces exhibiting symptoms of Fe deficiency mostly originate from relatively warm climates, such as India (37.5% accessions showing Fe deficiency) and Ethiopia (30%), where the pH is generally 6.5-7.0 This arose from either the chance introduction to these regions of Fe-deficient founder populations, or selection pressure in favour of either Fe-deficient types or genes linked to Fe-deficiency.

In a survey of patterns of morphological variation in the world germplasm collection of lentil, phenology was found as the key to the adaptation of the crop on a macrogeographic scale To further understand phenology, the flowering responses to temperature and photoperiod of a world collection of 369 accessions from 13 major lentil-producing countries (25 randomly selected accessions per country) together with lines from the ICARDA breeding programme were studied.

The distribution of country means for sensitivity to temperature and photoperiod illustrates the responses to selection for adaptation to new ecological environments following the spread of the crop from its origin Dissemination to lower latitudes such as into Egypt, Ethiopia and India was accompanied by a reduction in photoperiodic response.

Obligate photoperiodic control of the onset of flowering ensures that flowering starts annually in the same calendar period, irrespective of fluctuations in temperature. Consequently, selection against photoperiodic control in a long-day plant such as lentil implies an adaptation to relatively short days, which occur at low latitudes and which would otherwise delay flowering to an unacceptable extent.

Under these conditions, the crop relies rather more on temperature than photoperiod to ensure that flowering occurs at an ecologically and agronomically appropriate time. There was also evidence that flowering in the subtropical group is more temperature-sensitive than in West Asian germplasm, on average.

Movement from West Asia to higher latitudes, for example to Russia, resulted in a modest reduction in photoperiod sensitivity and an increase in temperature sensitivity, probably reflecting the change in sowing date from winter to spring.

Temperature sensitivity to processes other than flowering, such as the base temperature for rate of germination (T_b), also has been altered during dissemination from the Near East to environments with higher temperature regimes. The mean T_b of 15 randomly selected accessions from Ethiopia and India was higher than that of similar samples from Lebanon and Turkey

The degree of winterhardiness also has been affected in the spread of the lentil from its origins as a winter-sown crop in the Near East. A world collection of 3910 accessions was screened for winterhardiness near Ankara, Turkey, in the winter of 1979/80, when temperatures dropped to a low of -26.8°C and there were 47 days of snow cover In total, 238 accessions were undamaged by the cold winter.

Their origins were mostly from Chile (frequency of winterhardy accessions was 28%), Greece (33%), Syria (33%) and Turkey (15%), where selection for winterhardiness had occurred through winter sowing although lentil production is found in countries farther north, where the winter is colder, such as Russia (5%), and Hungary (4%), where sowing is exclusively done in spring. The frequency of winterhardy accessions from countries with warmer winter environments, such as Egypt, Ethiopia, India and Pakistan, was <1per cent.

The locus *aat-p* from the world collection show differences among countries and regions in the frequency of alleles and the extent of polymorphism, reflecting adaptation to local conditions There is evidence that a low frequency of *Aat-p*^{F} is associated with the spread into South Asia, whereas a higher frequency of *Aat-p*^{F} is characteristic of northern European material. Such regional differences result from the spread of the crop to new physical environments with the consequent natural and artificial selection for local adaptation.

To summarize with an extreme example, the spread of lentil from the Near East into India resulted in a loss of winterhardiness and the ability to extract iron from calcareous, high-pH soils, a reduction in photoperiodic sensitivity for flowering, increased intrinsic earliness in flowering, an increase in temperature sensitivity for flowering, and in the base temperature for rate of germination.

This list of factors affecting adaptation is not exhaustive. Each factor individually may be of minor importance, but collectively they illustrate the complex of interacting ecophysiological factors that determine adaptation. Superimposed on these natural selective forces is the effect of human selection on observable morphological traits, particularly the seed, for culinary and agronomic

The *pilosae* group of the Indian subcontinent is also characterized by two endemic qualitative morphological traits: precocity in flowering and maturity, and a low biomass. Furthermore, lentil germplasm from India is among the least variable among lentil-producing countries, despite India being the largest lentil-producing country in the world.

Recent evidence from isozyme (7 loci) and random amplified polymorphic DNA (RAPD) (22 loci) analysis indicates a similar striking difference between germplasm from South Asia and that from the rest of the world; additionally South Asian germplasm was found low in diversity germplasm.

The simplest explanation has the introduction around 2000 BC of a founder population adapted to the new environment that was probably both small in number and low in variability which resulted in a genetic bottleneck in South Asia (Erskine *et al.,* unpublished).

This still limits breeders' progress today. Indeed, a reconstruction of the phenological problems associated with the initial spread of the crop into the Indo-Gangetic Plain was inadvertently made with the introduction into that region of lentil selected in West Asia through ICARDA international nurseries in its early years Lentil selected in West Asia, when sown in India and Pakistan, mostly came into flower as the indigenous lentils were maturing.

Use of Information in Breeding

Armed with an understanding of the specific adaptation of lentil, the local constraints to production and the various consumer requirements of different geographic areas for seed, the breeding programme at ICARDA aims to produce genetic material suitable for national cooperators. The programme has been designed as a series of separate, but finely targeted streams linked closely to national breeding many of the selective forces important in the adaptation of the crop are environment-specific, such as temperature and photoperiodic sensitivities.

Clearly, selection undertaken in conditions very different from those in the target environment will have a lower response in the target environment than selection conducted directly within the target environment. Harnessing the specific comparative research advantages of ICARDA and its national programme partners, crosses (previously agreed with cooperators) are made and segregating generations advanced at ICARDA.

A major effort to widen the genetic base in the Indian subcontinent using three approaches, namely plant introduction, hybridization and mutation breeding. The plant introductions from West Asia flower as indigenous material matures in the Indian subcontinent.

The asynchrony in flowering has isolated *pilosae* lentils reproductively. However, the introduction of ILL 4605, an early flowering, large-seeded line, has resulted in its release as 'Manserha 89' for wetter areas of Pakistan and its widespread use as a parent in breeding programmes in the region. Hybridization between *pilosae* and exotic germplasm, primarily at ICARDA, followed by selection in the Indian subcontinent has resulted in cultivars with improved disease resistance and yield in Bangladesh and Pakistan.

The example of the widening of the genetic base of the lentil in South Asia illustrates the value of the historical perspective in plant improvement. Although ICARDA and its national programme partners are undertaking the first systematic lentil improvement programme with access to a wide range of genetic diversity and modern biotechnological tools, the efficiency of this research has been increased by an awareness that we belong to the approximately 200th generation of cultivators/selectors who have grown the crop since its domestication.

CROP EVOLUTION

Alphonse de Candolle, the founder of the study of crop evolution, was primarily interested in biogeography, the study of distribution of plants in relation to their environment. In his opinion, among the factors influencing this distribution are 'historic' factors such as glaciation events that are not related to the intrinsic adaptation characteristics of the species.

When he considered crop plants, he posited that one particular historic event - domestication - had had a paramount importance in determining their adaptation and, hence, distribution, as well as their genetic characteristics Four types of evidence could be utilized to determine the origin of domestication of crop plants These were archaeological (or more specifically archaeobotany), botanical (*i.e.* the distribution of the wild, ancestral relative), historical (or the existence of a written record documenting the existence or importance of the crop) and linguistic evidence (*i.e.* the existence of words designating the crop or objects or concepts related to the crop in native languages).

Of these four types, the first two provide the most abundant and reliable evidence. Since the time of de Candolle, however, additional scientific methods have increased considerably our power to determine centers of crop domestication based on archaeological and botanical arguments.

Although common bean is not a species originating in the Near East, the focus of the present symposium, it presents nevertheless an interesting domestication history, which is only recently being elucidated.

The first botanically accurate description of common bean in European herbalsWhereas barely 50 years had passed after the conquest of the Americas, the actual origin of the crop had already been lost to the writer of the herbal and later European botanists. Linnaeus assigned the origin of common bean to India; de Candolle himself expressed doubts that common bean had been domesticated in the Americas.

It was only during and after the Second World War that it became generally admitted that common bean originated in the Americas. The conclusion was based on the discovery of wild common bean in Argentina and Guatemala and archaeological remains in the Americas.

Based on the extensive distribution of the wild relative speculated that common bean originated through multiple domestications as did other crops originating in the Americas, such as cotton, pepper and amaranth.

Observed that there was a Mesoamerican cultivar group with smaller seeds and an Andean group with larger seeds. None of these observations, however, provided substantial experimental evidence and hence a clearer picture of the actual process of domestication for this species.

For example, one could ask whether the two cultivated groups identified resulted from distinct domestications or a single domestication event followed by divergence into a small-seeded and a large-seeded group. One could also ask where these single or multiple domestications took place. The archaeological record of the origins of agriculture in the Americas is rather limited.

In Mesoamerica, This important type of information therefore pales in comparison with what is available in the Near East In part, this may be due to the relatively humid environment required by many crop relatives in the Americas, which render long-term conservation of crop remains less likely. Therefore, in relative terms compared with archaeological information, there is a relatively higher emphasis on botanical evidence to decipher crop evolution in the Americas than in the Near East.

The use of molecular marker information has increased our understanding of the domestication process in common bean a crop originating in the Americas. In addition, I hope to demonstrate how molecular analyses can bridge the gap that has separated botanical and archaeological evidence, the twin pillars on which rests the study of crop evolution. The power of molecular biology to change scientific paradigms is by no means unique to this field but extends to many other fields as well.

GENERAL METHODOLOGICAL PRINCIPLES

Our current understanding of the origins of domestication in common bean is based on a two-pronged approach relying on the study of wild-growing *P. vulgaris* beans, on one hand, and the use of biochemical or molecular markers, on the other.

The use of wild beans is predicated on the fact that their dispersal occurs over much smaller distances (at most a few meters after each vegetation cycle, if at all) than their domesticated descendants. The latter are subject to long-distance exchange, trade, gift or human migration, all of which tend to obscure the original geographic distribution patterns of genetic diversity.

Recent explorations have provided a more complete representation of the actual distribution of wild *P.* beans are distributed from approximately 30°N Lat. to 30°S Lat. at mid-elevations from 1000-2500 m asl in regions with moderate rainfall. A common characteristic of the environments in which wild beans are found is the existence of a marked dry season, an important feature needed for seed dispersal.

Genetic relatedness is traditionally assessed with morphological phenotypic traits. Plants with a similar phenotype are inferred to be genetically related either through common ancestry or through geneflow (hybridization). Occasionally, however, inferences based on phenotypic similarity are erroneous. Plants can arise from different selection episodes leading to the same phenotype (convergent evolution).

In addition, similar phenotypes can have a different genetic basis (genetic heterogeneity). In both cases, an inference of genetic relatedness would be unwarranted. In contrast, biochemical, and even more so, molecular markers provide information on genetic relatedness that is generally (although not always) devoid of selective influences.

All biochemical and molecular markers were created equal! Differences have been observed with regard to the ease of analysis, reproducibility, level of polymorphism, number and genome distribution of loci Of particular note is the molecular basis of polymorphisms.

The more complex the basis, the more desirable the marker, because complexity will reduce the probability of repeat mutations, especially in conspecific materials, which are, by definition, closely related. For example, polymorphisms based on DNA nucleotide substitutions are simple changes that could presumably occur repeatedly.

Analyses based on single or few RAPD or RFLP markers are therefore unreliable in determining genetic relatedness that sample a large number of changes at the molecular level, because they include either a large number of markers or complex markers such as actual DNA sequences or multigene families, are more robust with regard to repeated changes.

For example, banding patterns of seed proteins such as phaseolin in common bean, when analyzed in denaturing SDS-polyacrylamide gel electrophoresis, represent a compomy image resulting from changes at the DNA nucleotide level (duplication of genes, within-gene duplications, nucleotide substitutions) and at the translation level (co - and post-translational modifications).

Therefore each pattern is unlikely to have arisen more than once. As a consequence, genotypes that exhibit the same protein pattern are likely to share a common ancestor

One of the major experimental difficulties in crop evolutionary studies is the distinction one has to make between ancestry (domestication) and geneflow as a cause of genetic relatedness.

This is particularly the case given that wild ancestors and domesticated descendants belong to the same biological species and can therefore freely hybridize.

Hence, similarity between ancestors and descendants even at the molecular level cannot be assumed to be due to domestication For example, DNA sequence data and the utilization of an outgroup (*e.g.* closely related species) can shed additional light on this problem because they can provide information on the nature of the polymorphism and the direction of the change, *i.e.* the ancestral vs. descendant states of the molecular polymorphism.

Domestication Pattern

The origin of the *Phaseolus vulgaris* progenitor in its wild state can be traced to the Pacific slope of Ecuador and northern Peru, based on the distribution pattern of direct repeats in phaseolin seed protein genes Phaseolin is coded by a small multigene family inherited as a single Mendelian unit and located on linkage group B7 The phaseolin locus is at most 190 Kb long and contains 6 to 9 genes Differences among these phaseolin sequences include nucleotide substitutions and the presence or absence of tandem direct repeats.

Three repeats have been identified so far: a 21 bp repeat in the third intron, a 15 bp repeat in the fourth exon and a 27 bp repeat in the sixth exon. The existence of these tandem repeats provides an interesting landmark to follow the evolution of phaseolin genes. Indeed, the generation of a repeat is a more likely event than the precise loss of such a repeat, which would be needed to restore a sequence without repeats. Hence, phaseolin genes without repeats represent probably a more ancestral state than phaseolin genes with repeats. This was verified by analyzing the presence of tandem repeats with a polymerase chain reaction (PCR) test of phaseolin sequences in two subspecies of a related species, *P. coccineus* subsp. *polyanthus* and *P. coccineus* subsp. *coccineus.*

In these two taxa, none of the three repeats could be identified, suggesting that the repeatless state is indeed the ancestral state. A survey was then conducted among wild *P. vulgaris* to determine if any phaseolin haplotypes lacked genes with tandem repeats.

Only wild bean populations from Ecuador and northern Peru on the Pacific slope of the Andes appeared to be devoid of phaseolin genes with tandem repeats It was inferred that these wild populations therefore represent the presumed ancestor of the species. This conclusion is somewhat unexpected as most wild *Phaseolus* species are distributed in Mexico.

One could have thought that *P. vulgaris* would have originated in that region as well and that other wild *P. vulgaris* populations distributed elsewhere were ultimately derived from Mexican ancestors. It may be, however, that some wild *Phaseolus* ancestors ('proto-vulgaris' and other species) were dispersed from Mexico to Central America and the Andes to give rise to locally distributed wild *Phaseolus* species, such as *P. costaricensis* From such a local distribution, a species such as *P. vulgaris* could have acquired a secondary, broader distribution by dispersal towards the north and south. It is worth remembering that lima bean (*P. lunatus*) also shows an extensive distribution of its wild relative from Mexico to Argentina and that the species consists of two genepools, an Andean genepool originating on the Pacific side of the Andes in Ecuador and northern Peru (although at lower altitudes than wild *P. vulgaris*) and a 'Mesoamerican' genepool distributed from Mexico to the southern Andes. Although the domestication region of the 'Mesoamerican' genepool is unknown so far, it is likely to be located somewhere in Mesoamerica, given that most Mesoamerican cultivars are located there

From the small nuclear area in Ecuador and northern Peru, wild *P. vulgaris* were dispersed towards both the north (Colombia and Venezuela, Central America and Mexico) and the south (southern Peru, Bolivia and Argentina) to achieve their current distribution In a subsequent step, multiple domestications took place in the Mesoamerican and Andean regions.

At least one domestication may have taken place in west-central Mexico (Jalisco), a conclusion based on phaseolin data. Most Mesoamerican cultivars exhibit the same phaseolin type ('S'-type).

Wild beans with the same phaseolin type and without morphological signs of past hybridizations to domesticated beans are concentrated in the Mexican state of Jalisco, Interestingly, Doebley *et al.* suggested a similar domestication area for maize.

A second major domestication took place in the southern Andes although a more specific region awaits confirmatory. Phaseolin suggest possible multiple domestications within the Andean region because at least four phaseolin types have been identified among Andean beans.

Alternatively, the original populations that were domesticated were polymorphic for phaseolin. Further are needed to distinguish between these possibilities. A third, likely minor domestication may have taken place in Colombia, although regions in Central America deserve further study because samples of wild *P. vulgaris* from that area were unavailable.

POST-DOMESTICATION DIVERGENCE

Common bean is perhaps best known for the high level of phenotypic diversity, particularly for seed type (colour, colour pattern, size and shape). This broad diversity is a consequence of the wide range of cultural environments under which common bean is grown, both in the Americas and elsewhere.

This diversity has made it difficult to recognize patterns of genetic diversity in the domesticated genepool because of the possibility of convergent and reticulated evolution.

Analyses of biochemical traits such as isozymes and phaseolin seed protein have provided additional information on this problem. Koenig and Gepts and Singh *et al.* analyzed isozyme diversity in wild and domesticated germplasm, respectively. From these studies, it was possible to identify clusters of cultivars within the Andean and Mesoamerican genepools that shared a common isozyme allele. Further analyses of phenotypic diversity were conducted with multivariate statistical analyses (canonical and discriminate analyses) by morphological, agronomic and ecological traits.

For example, in the Mesoamerican genepool, three subdivisions or races can be distinguished Race Mesoamerica represents cultivars originating predominantly from the warmer, more humid areas of Mexico, Central America and Colombia. Possibly, this race could be split further to account for an additional, minor domestication area in Central America or Colombia

Race Durango includes cultivars from the northern highlands of Mexico, which have a cooler, more arid environment. Race Jalisco consists of the most traditional common bean cultivars, originating predominantly in the humid central and southern highlands of Mexico.

A similar subdivision in three ecogeographical races has been established for the current status of the domestication scenario in common bean. process The arrows indicate rare cases of geneflow between the two major genepools.

DOMESTICATED BEAN

Wild relatives and their crop descendants show marked phenotypic differences, collectively called the domestication syndrome As a consequence, they have been classified in different taxonomic species, which is unjustified given that often they can be freely crossed. Based on the traditional neo-Darwinian theory of evolution, one would expect that these marked phenotypic differences would be controlled by a large number of genes, each with a small phenotypic effect. This view reflects a view of evolution involving very gradual changes in environment and adaptation. This may not have been the case during domestication (nor in natural environments for that matter) where the domesticated and natural environments are quite different. Hence, adaptation to cultivation represented selection for a radically different environment over a time scale that would have been very short in an evolutionary context (even a few 100 generations should be considered a very short period in evolution). This suggested that major genes might have played a role during domestication because only such genes would have provided a fast response to the rapidly changing environment imposed by cultivation. To test this view of evolution during domestication, an experiment was conducted to determine the inheritance of the domestication syndrome in common bean A recombinant inbred population was established from a cross between 'Midas' (a snapbean cultivar representing the 'ultimate' in bean domestication) and 'G12873' (a wild bean from Mexico). In this RI population, a linkage map was constructed with RFLP markers regularly spaced throughout the genome. In turn, this map was used to conduct a quantitative trait locus (QTL) analysis for quantitative traits and to map major genes involved in the domestication syndrome. The following traits were studied.

POD DEHISCENCE

The onset of the dry season coincides with the maturation phase of the life cycle of wild beans. During the progressive desiccation of the plant, fibres located in the pod walls and sutures shorten and, hence, cause the explosive dehiscence of the pods. (Without desiccation during the dry season, pods rarely open and seeds germinate in the pods, causing a marked reduction in fitness.) Pods of domesticated beans contain fewer or no fibres and therefore do not open at maturity. This difference in the capability of seed dispersal is one of the most important characteristics distinguishing wild and domesticated wild beans. It can be easily recognized in dry pod remains: dried wild pods are strongly twisted (owing to the oblique orientation of fibres in the pod walls), whereas dried domesticated pods are not or only slightly twisted. Presence or absence of fibres is controlled by the *St* gene.

Seed Dormancy

Because the onset of the growing (wet) season is irregular, plants have adopted seed-dormancy mechanisms to insure against premature germination during the first rains. In common bean, dormancy is essentially impermeability of the seed coat, which prevents immediate imbibition of the seed. Prior to this study, no studies had ever been undertaken about the inheritance of dormancy in common bean.

Growth Habit

Wild beans are viny plants that use shrubs or trees as support. During common bean evolution, there has been a selection for a bush growth habit. Growth habit consists of several subtraits, including the number of nodes and pods on the main stem, the length of the internodes, and type of stem termination (vegetative or indeterminate vs. reproductive or determinate). The latter traits are conditioned by the *fin* gene.

Photoperiod Sensitivity

Photoperiod sensitivity plays a crucial role in scheduling of flowering, such that maturation will take place at the beginning of the dry season. Earlier maturation will result in reduced or lack of seed dispersal.

Later maturation will prevent sufficient development of seeds before the dry season sets in. Common bean, having originated in the tropics, is a short-day (long-night) plant, which will flower only under daylengths of 12 hours or less. During dispersal from centers of domestication, dayleingth neutrality was selected, whereby common bean plants will flower under any daylength. Photoperiod sensitivity is controlled by the *Ppd* gene.

Seed Size and Colour

Selection by farmers and consumers has led to larger pods and seeds with different colours, among them white (absence of pigmentation) conditioned by the *p* gene. Four conclusions were drawn from the results.

First, for many traits it was possible to identify major genes. For example, at least four genes controlled seed dormancy, one of which accounted for at least 50% of the phenotypic variation.

Second, for most of the traits, more than 50% of the phenotypic variation could be accounted for in genetic terms, suggesting that heritability of most traits was high.

Third, distribution of genes involved in the domestication syndrome appeared to be non-random, as there was a relatively higher concentration of genes on linkage groups 1, 2 and 7. Linkage group 1 appeared to be particularly important because it included genes for photoperiod (*Ppd*), determinacy (*fin*) and internode length.

CULTIVATED CEREAL CROPS

The crucial difference between cultivated cereal crops and their wild relatives is in ear fragility at maturity. Easy shattering of spikes into spikelets upon maturity is essential for seed dispersal and survival in the wild, whereas forms with non-brittle ears survive only under cultivation.

It is generally assumed that most Triticeae crops have been domesticated from their wild relatives by selection of non-shattering individuals which sporadically appear in wild populations as rare mutants.

However, there are no documented cases of the appearance of non-brittle mutants in wild populations, except observations of introduction from cultivars into weedy forms.

The appearance of non-brittle mutants seems to be a rare event which may be induced under specific conditions. Interspecific hybridization may activate transposition of mobile genetic elements which may be one potential source of increased mutation rates.

A hypothesis was proposed that rye and wheat forms with varying ear fragility may have arisen as a result of interspecific hybridization processes between different wild species (Jaaska 1975).

Taking into account that non-brittle mutants will persist in the wild only for a limited number of generations after their appearance, it follows that rye and barley should have been domesticated *in statu nascendi* from the ancient hybrid populations of their wild relatives.

IMPORTANT CULTIVATED TREES OF OLIVE

Olive (*Olea europaea* L.) is one of the most important cultivated trees in the Mediterranean basin and the spread of its cultivation was associated with that of grapevine, These species form the oldest group of domesticated trees. *Olea europaea* is a complex of four subspecies *Olea europaea* subsp. *europaea* is separated into two taxa: var. *sylvestris* (oleasters: feral and wild forms) and var. *sativa* (cultivated forms). Cultivated trees could develop from oleasters.

The area of olive domestication appears to have been the Near East. Artifacts associated with the oil-extraction process as well as those associated with clonal propagation are the main evidence for domestication which occurred during the early Bronze Age.

Nevertheless, the exploitation of wild forms before domestication is suggested by discoveries of olive This was not specific to the Near East. On the basis of archaeobotanical evidence, Terral and Arnold-Simmard have suggested a pre-domestication period in eastern Spain and Palestine.

During this period, harvesting and caring of wild trees could have occurred without vegetative multiplication, thus leading to higher production. Evidence of olive remains from the Holocene, outside the natural distribution of the species, but the low frequency of olive fragments is insufficient to conclude that domestication occurred there.

Table. The *Olea europaea* L. complex: Subspecies and their Main Taxa.

Subspecies		**Geographical origin**
subsp. *europaea* Mansfield		
var. *sylvestris* D.C.	- *O. europaea L.* Mediterranean basin (oleasters)	
(cultivars)	var. *sativa* Lehr.	Mediterranean basin
subsp. *laperrinei* Ciferri		
	- *O. laperrinei* Batt. & Trab.	Saharan Mountains
& Burd.	- *O. maroccana* Gren. (Morocco)	Atlas Mountains
subsp. *cerasiformis* Islands	Canary and Madeira	Kunk & Sund.
subsp. *cuspidata* Ciferri		
	- *O. cuspidata* Wall.	Asia
	- *O. chrysophylla* Lam.	Arabia, Abyssinia
	- *O. africana* Mill. Africa	Eastern and southern

The number of generations required for the improvement and selection of cultivated forms is likely to be very low in comparison with annual plants. A close genetic similarity is therefore expected between wild forms and their direct cultivated descendants has suggested that cultivated forms were propagated in the entire Mediterranean basin and crosses between them and with local wild forms have led to the creation of new cultivars.

At least, the crosses between related species of the *Olea europaea* complex could have occurred in different areas (northern Africa, the Middle East and the Canary Islands).

In southern Morocco, *O. maroccana* has been considered as an intermediate form between *O. europaea* and *O*The distribution area of this form could overlap with the cultivated form Some examples of spontaneous interspecific crosses are known for several cultivated trees, notably apple, pear, date palm and grapevine Consequently, the origin of Mediterranean olive is not clear. Two places have been suggested: North Africa and the Near East

A molecular analysis of different wild and cultivated accessions could clarify the origins of the domestication and of the selection of olive trees.

The RAPD markers have been used to identify cultivars to reveal the extent of diversity and to differentiate species for detection of interspecific hybridization or in a phylogenetic In this preliminary study we chose cultivars and wild forms with RAPDs to estimate the similarities between them.

MATERIAL AND METHODS

Sixty-four cultivars from different countries around the Mediterranean basin were analyzed These cultivars were selected from a germplasm collection maintained at Montpellier (France) and provided by Dr Rallo (Cordoba) and Dr Baldoni (Perugia) from the collections of Cordoba (Spain) and Perugia (Italy).

Oleasters (wild olive trees) were studied from three sources: from Corsica Morocco (9 individuals: 'Morocco 1' to '9'), and from southern France We have no criteria to distinguish wild from feral forms. Four individuals of *O. maroccana* from southern Morocco and four individuals of *O. europaea* subsp. *cuspidata* from four different areas were also studied.

DNA preparation was carried out modified method described by Saghai-Maroof *et al*Five grams of leaves were ground and 20 ml of 2X CTAB bb1. This mixture was centrifuged at 4500 rpm for 20 min. The supernatant was recovered and 10 ml of chloroform/IAA(24/1) were added and mixed at 100 rpm for 20 min.

Table. List of *Olea* Cultivars to their Geographic Origins.

Geographical Origin	No. of cvs.	Cultivars
Maghreb	9	Barouni, Bid el Hamam, Chemlal de Kabylie, Chemlali de Sfax, Chetoui, Meski, Picholine Marocaine, Sigoise, Zarazzi
Iberian Peninsula	8	Arbequina, Cornicabra, Empeltre, Galega, Lechin de Sevilla, Manzanilla, Picual, Villalonga
France	20	Amellau, Belgentier, Bouteillan, Cayet Bleu, Cayet Rouge, Cayon, Colombale, Corniale, Filitosa[†], Sabina, Lucques, Olivière, Picholine, Picholine de Rochefort, Pigale, Rougette de Pignan, Tanche, Verdale, Verdelet, Vermillau
Italy	11	Ascolana Tenera, Cassanese, Cellina, Dolce Aggogia, Frantoio, Giarraffa, Leccino, Leucocarpa, Nocelara del Belice, Pendolino, San Felice
Yugoslavia	1	Oblitza
Greece	6	Amygdalolia, Carolia, Gaïdourolia, Kalamata, Koroneiki, Valanolia
Turkey	4	Ayvalik, Domat, Sofralik, Uslu
Near East	4	Kaissy, Merhavia, Souri, Zaity
Egypt	1	Toffahi
Total	**64**	

This mixture was centrifuged at 4500 rpm for 20 min. The supernatant was recovered and mixed with 20 ml of isopropanol and stored at -20°C overnight. The DNA pelot was recovered, washed with ethanol solution (ethanol 76%, 10 mM ammonium acetate), dried, and removed in 0.5 ml of 1X TE (10 mM Tris-HCl pH 8,1 mM EDTA).

Then 500 μg of RNAse A (Boehringer) were added. The incubation was performed at 37°C for 1 hour; 530 μl of phenol/chloroform/IAA (25/24/1) were added and mixed. After centrifugation at 12,000 rpm for 10 min, the supernatant was recovered and mixed with 530 μl of chloroform/IAA (24/1).

Another centrifugation was performed at 12,000 rpm for 10 min, and then the supernatant was mixed with 1 ml of ethanol and 188 μl of 7.5M ammonium acetate solution. This mixture was kept overnight at -20°C. A final centrifugation at 14,000 rpm for 10 min was performed to recover DNA pelot which was dried, then dissolved in 0.5 ml of 1XTE.

The procedure for obtaining RAPD markers has been described by Quillet *et al.* Thirty decamer primers (Bioprobe) were tested on eight genotypes: five cultivars ('Olivière', 'Lucques', 'Giarraffa', 'Arbequina', 'Domat'), two wild forms corresponding to the *cuspidata* subspecies (from Iran and China) and *Ligustrum vulgaris* L. as an outgroup. The choice of primers was based on the level of polymorphism with the clearness of banding profiles.

RYE

Despite considerable disagreement among taxonomists about the delimitation of rye species and their intraspecific taxa the rye genus *Secale* L. may be treated as consisting of three valid biological species

- The Outcrossing annual *Secale cereale* L. s.l, including the European cultivated rye (subsp. *cereale*), the Transcaucasian cultivated and weedy rye with a tough rachis (subsp. *segetale* Zhuk.), and various weedy and wild forms of the annual self-incompatible rye with a brittle rachis such as subsp. *ancestrale* Zhuk. s.l, including weedy subsp, *ancestrale* Zhuk. s.str, weedy subsp. *afghanicum* (Vav.) Hammer, weedy subsp. *dighoricum* Vav., and wild subsp. *vavilovii* (Grossh.) Kobyl. s.str., non s.l. of Kobyljanskij.
- The Outcrossing perennial *Secale strictum* Presl., syn. *S. montanum* Guss., with several subspecies: subsp. *strictum* (= *S. dalmaticum* Vis.), subsp. *anatolicum* (Boiss.) Hammer, subsp. *kuprijanovii* (Grossh.) Hammer, subsp. *ciliatoglume* (Boiss.) Hammer, and subsp. *africanum* (Stapf) Hammer.
- *Secale Sylvestre* Host. - a cleistogamous annual of sandy dunes and coasts.

A cleistogamous annual, originally collected by Kuckuck in Iran and attributed to *S. vavilovii* may deserve a specific rank as *S. iranicum* Kobyl., syn. *S. vavilovii* Grossh. sensu Khush, non sensu Grossheim or sensu lato of Gandilyan owing to its reproductive isolation as a result of its cleistogamous breeding system. Likewise, South African endemic perennial rye may be recognized as *S. africanum* Stapf because of its autogamous breeding system.

Grossheim described *S. vavilovii* as a short-statured (stems 20-35 cm) wild annual with short (4-8 cm) fragile spikes which he collected in Nakhitshevan from a dry habitat on volcanic ash soil. Populations of short wild annual rye were later found to occur in different dry regions of Armenia by Gandilyan who attributed the short stature to a phenotypic adaptation to a dry and nutrient-poor habitat.

On those grounds he proposed to treat all wild and weedy annual ryes with a brittle rachis as *S. vavilovii* s.l. separately from the cultivated *S. cereale* s.str. To give these forms a specific rank is questionable because they are not reproductively isolated from non-brittle weedy *S. cereale* subsp. *segetale.* Their inclusion under *S. cereale* subsp. *vavilovii* s.l. as proposed by Kobyljanskij contradicts the botanical nomenclature rules because of the priority of subsp. *ancestrale* Zhuk.

At the subspecies level Vavilov expressed a view that cultivated rye evolved from a wild perennial rye through the appearance of annual weedy forms which in turn were domesticated by unconscious selection of non-brittle forms under cultivation in barley and wheat fields. Later he reported finding a weedy rye in Afghanistan with fragile ears which he attributed to *S. cereale* var. *afghanicum* and proposed that it might be an initial form from which semi-brittle and non-brittle forms of weedy rye were spontaneously selected under cultivation in wheat and barley fields in northern areas and higher altitudes He also further argued that *S. cereale* may have evolved from a perennial *S. montanum* with wild *S. vavilovii* Grossh. acting as a link between them. rye was domesticated in the Near East region as a secondary weedy crop (subsp. *segetale*) and only thereafter spread to European countries and was used as a distinct crop.

Alternatively, the present-day weedy ryes with brittle and semi-brittle rachis are products of introgressive hybridization of the cultivated non-brittle subsp. *segetale* with wild rye subsp. *vavilovii* s.str. Segregants from this hybridization acquired a weedy trait from subsp. *segetale* and became field weeds.

Artificial hybrids between wild, weedy and cultivated forms of *S. cereale* are completely interfertile and show regular meiosis (that they comprise a single biological species.

It is generally accepted that all annual rye species have evolved from the perennial *S. strictum*. *Secale cereale* and *S. strictum* differ by two large translocations between three of the seven chromosomes, but intraspecific taxa have the same chromosome arrangement in both species Polymorphism for reciprocal translocations has been described in populations of *S. cereale.*

The evolution of *S. cereale* from *S. strictum* should thus involve fixation of two reciprocal translocations which would be possible only by the involvement of an autogamous intermediate. Stutz (suggested that cleistogamous annuals *S. sylvestre* and *S. vavilovii* sensu Khush (= *S. iranicum*) might represent intermediate stages in the evolution of *S. cereale* from *S. strictum.*

Secale sylvestre has the same chromosomal arrangement as *S. strictum,* but a low crossability with it, or differs in only a small chromosomal translocation, whereas *S. iranicum* was found to have the same chromosomal arrangement as *S. cereale* (Nürnberg-Krüger Stutz suggested that cleistogamous annual *S. sylvestre* has been derived from *S. strictum* and has then given rise to a similarly cleistogamous annual *S. iranicum* by fixation of two chromosomal translocations, while *S. cereale* was assumed to emerge from *S. iranicum* by acquisition of the outcrossing breeding system by introgression from *S. strictum.*

Variability among the rye species established that *S. cereale* and *S. strictum* display partially homologous polymorphism of many isozymes with most allozymes shared between them. that *S. iranicum* is fixed for one of the allozymes found in *S. cereale* and *S. strictum,* whereas *S. sylvestre* has unique allozymes of acid phosphatase ACP-B, anodal peroxidase PRX-C, and aromatic alcohol dehydrogenases AAD-A and AAD-E In the it has been concluded that *S. sylvestre* is a lateral branch of evolutionary divergence form S. *strictum* and not an intermediate leading to *S. iranicum* and *S. cereale.*

Morever, *S. strictum* and *S. sylvestre* are well isolated in nature by growing in different ecological habitats and do not hybridize freely. Instead, I suggest that some unknown inbreeding form of subsp. *vavilovii* related to *S. iranicum* might be an intermediate from which both brittle- and tough-rachis forms of *S. cereale* evolved by introgression of self-incompatibility from *S. strictum* It was also assumed that non-brittle rye *S. cereale* subsp. *segetale* might be domesticated *in statu nascendi,* directly from some such hybrid populations currently extinct, while fragile subsp. *vavilovii* s.str. has persisted in the wild. Later, *S. sylvestre* was found to differ from the other rye species also in allozymes of cathodal peroxidases CPX-4 and CPX-5 and chloroplast DNA whereas *S. cereale* and *S. strictum* revealed homologous polymorphism and could not be definitely distinguished by allozymes and chloroplast DNA

Zhukovsky (1pointed out that there have been no reports of spontaneous appearance of non-brittle forms for more than 100 years of cultivation of wild and weedy ryes in different botanical gardens and collections. The reason may be in the recessive nature of the non-brittle mutation combined with the outcrossing breeding system of rye The rare non-brittle mutants will remain phenotypically unexpressed in outcrossing rye populations in heterozygous genotypes until two parents with recessive non-brittle alleles cross and segregate into homozygotes in a progeny. The probability for this depends on the frequency of non-brittle mutations which may be extremely low. Therefore, the involvement of an inbreeding intermediate would be an important premise for the appearance of recessive non-brittle mutants phenotypically expressed in homozygotes.

Zhukovsky also has proposed a hypothesis that weedy subsp. *segetale* might have arisen from interspecific hybridization between wild perennial and annual ryes with the appearance of non-brittle forms in the hybrid progeny. He assumed that interspecific hybridization might be the crucial mutagenic factor which has caused the appearance of semi- and non-brittle forms of rye. Stutz "on the slopes of Mt. Ararat, S. *vavilovii* makes contact and hybridizes rather freely with *S. montanum.*"

The seed obtained from crosses between *S. cereale* and *S. strictum* were found to germinate only when *S. cereale* was the mother parent indicating that the introgression is possible from the perennial rye. Sympatric populations of wild perennial and annual ryes in the Near East region should be examined for the fragility, chromosome rearrangements and molecular characters in order to have more information about the origin of wild, weedy and cultivated forms of *S. cereale.*

Electrophoretic variants of aliphatic and aromatic alcohol dehydrogenase (ADH and AAD), aspartate aminotransferase (AAT), acid phosphatase (ACP), esterase (EST) and anodal peroxidase (APX) isoenzymes in rye species: major electrophoretic variants (allozymes) are numbered in the order of their decreasing mobilities and listed in the order of decreasing occurrence; rare allozymes are labelled by letter 'r'; unique allozymes are in italics.

Species	**ADH**	**AAD**	**AAD**	**AAT**	**AAT**	**ACP-**	**EST**	**EST**	**APX**
	-A	**-A**	**-E**	**-B**	**-C**	**B**	**A**	**B**	**C**
S. strictum	2;1r	1	2;1;3r	1	2;1	4;4f;5;3	2f;2;3	3;1;2	1
S. cereale									
subsp. cereale	2	1	2;1	1	2;3	4;1;3;2	2	1	1
subsp. segetale	2	1	2,1	1	2;3	4;1;3;2	2	1	1
subsp. ancestral	2	1	2;1	1;2r	2;3	4;3;1;2	2;2f;3r	1;3;2	1
S. iranicum	2	1	1	1	2	4	2	1	1
S. sylvestre	2	*2*	*1f*	1	2	*4f*	2;0	1	2

BARLEY

The cultivated barley *Hordeum vulgare* L., incl. *H. distichon* L. and *H. hexastichon* L., and its closest wild relative *H. spontaneum* C. Koch are autogamous annuals which with the allogamous perennial *H. bulbosum* L. share basic genome a number of allozymes Their hybrids are completely interfertile and have normal chromosome pairing in meiosis. Cultivated barley and its closest wild relative, *H. spontaneum,* are now commonly recognized as subspecies of *H. vulgare* s.l.

However, taking into account their autogamous breeding system, which itself provides a reproductive barrier between them, and the fact that the biological species concept is not strictly applicable to uniparentals, they could formally be accepted at a species level as separate phylogenetic lineages.

There is a general agreement among investigators that cultivated barley has originated from its closest wild relative *H. spontaneum* through the latter's domestication Different views, however, have been expressed with respect to the place, time and mechanisms of the origin of different forms of cultivated barley, particularly six-row forms. Monophyletic origin of the cultivated barley as a two-row form by domestication of *H. spontaneum* and the secondary origin of the six-row barley from a cultivated two-row form is now generally accepted thanks to experimental evidence on the secondary origin of the Tibetan semi-brittle six-row barley by hybridization between the cultivated six-row barley and wild *H. spontaneum.*

The semi-brittle forms in the present-day mixed populations of weedy and cultivated barley are evidently derived by introgression from the cultivated barley. Genetic studies of hybrid progeny between *H. vulgare* and *H. spontaneum* have shown that ear fragility is controlled by two closely linked genes (*Bt* and *Bt2*) with a recessive mutation at either loci giving rise to non-brittle ears In the case of a simple *in statu nascendi* domestication of barley by selecting out non-brittle mutants, the cultivated barley has evidently been domesticated repeatedly from different genotypes of wild barley, *i.e.* it has a polytopic origin.

This, however, does not explain a strikingly wider morphological variability among cultivated barley than in other cereals. Indeed, morphological variation within *H. spontaneum* is limited to only three botanical varieties which differ in the apex shape and awn length of lateral spikelets.

At the same time, the cultivated barley is remarkably more variable with over 100 botanical varieties described among primitive barley landraces Although a remarkable morphometric variation was observed among a set of 77 accessions of wild barley under cultivation around Moscow (Russia) under unusual climatic conditions, it still remained relatively limited and no non-brittle cultivar-like forms were observed Characters of the wild parent dominate in hybrids with cultivated barley and segregation appears in later generations.

No morphologically new types were found among the progeny of hybrid generations between various morphological types of *H. spontaneum.* Therefore, the question as to why cultivated barley is so morphologically variable still remains unanswered.Comparative studies of isoenzyme variation among a set of morphologically different accessions of *Hordeum vulgare* s.str. and *H. spontaneum* showed that both-had the same allozymes of essentially monomorphic isoenzymes and displayed variation of 10 polymorphic esterase isozymes.

This shows that the isozyme loci which were monomorphic in wild barley have remained also in the cultivated barley, *i.e.* cultivation has not induced new variation at the isozyme loci in contrast to those controlling variable morphological characters. The same also has been demonstrated with respect to wild and cultivated wheats and The cultivated barley and its wild progenitor could not definitely be distinguished by any of the isoenzymes studied except by a frequency of allozymes of some isoesterases.

Hordeum bulbosum displays extensive intrapopulation polymorphism of several isozymes which were largely monomorphic in *H. vulgare* and *H. spontaneum.* The allozyme genepool of the two annuals was found to be a subset of most frequent allozymes of *H. bulbosum* This result is consistent with a view that *H. spontaneum* might have evolved from *H. bulbosum* by fixation of genes controlling self-compatibility and annual habit, polytopic origin of *H. spontaneum,* as well as against later, repeated introgression from the bulbous barley. Moreover, crossing experiments between *H. vulgare* s.l. and *H. bulbosum* have shown that hybrids could be obtained only by the embryo rescue technique, had disturbed meiosis, were sterile and frequently haploid owing to spontaneous elimination of *bulbosum* chromosomes.

Isozyme	**Vulgare**	**Spontaneum**	**Bulbosum**	**Vulgare**	**Spontaneum**
ADH-A		1	1	1;2r	
ADH-B	1	1	1		
ADH-D	3	3	3;2;1		
AAD-A	1	3	3;4;2;1		
AAD-B	2;0	2;0	2;2s;1;0		
AAD-E	2	2	2;1;3		
AAT-A	1	1;2	1		
AAT-B	1	1	1;2;3;0		
AAT-C	2	2	2;3;1		
EST-A				1	1
EST-B				1	1
EST-D				3;2;4;0	2;3;4;1;0
EST-E				2	2;1r;3r
EST-F				1;2;0	2;1;3;4
EST-H				2;0	2;1;0;3
EST-I				2;0;1;3	2;0;3;1
EST-J				1;0	1;0
EST-K				3;0;2	0;1

Electrophoretic variants of aliphatic and aromatic alcohol dehydrogenase (ADH and AAD), aspartate aminotransferase (AAT) and esterase (EST) isoenzymes in cultivated barley (*vulgare*) and its wild relative species (*spontaneum, bulbosum*): major electrophoretic variants (allozymes) are numbered in the order of their decreasing mobilities and listed in the order of decreasing occurrence; rare variants are labelled by letter 'r'.

Hordeum spontaneum and *H. bulbosum* are reproductively isolated by strong sterility barriers. Observations that crossability and chromosome elimination varies depending on the genotype of parental species suggested that introgression from the bulbous barley in some genotype combinations may be possible.

OTHER CROPS

The diversity of barley (*Hordeum vulgare*) is more evenly spread across the Ancient Mediterranean area and several infraspecific taxa are endemic to Ethiopia The botanical varieties of the genus *Pisum* are also more or less evenly spread in the Ancient Mediterranean area of origin Several monographs dealing with many different crops have been published by the VIR.

Such work is important not only for studies on centers of origin of cultivated plants but also for theoretical and practical agronomy. Such work is the basis for searching for initial material for plant breeding.

Taxonomic studies in the tradition of Vavilov, based on the investigation of variation within a species, have also been carried out in Gatersleben Recently information on coriander has been provided by Diederichsen The detailed investigation of the variation of the species *Coriandrum sativum* by several characters caused the author to divide this species into several groups (so far called ecological types), which are connected with the geographical origin in the Ancient Mediterranean area.

The infraspecific classification helps to indicate areas where different types of a given species are to be found. It also is an excellent method to single out and preserve rare accessions in a collection. However in the case of wheat a very interesting group - *T. durum* convar. *villosum* - collected by Vavilov in Syria, Jordan and Lebanon is untraceable in the collection at VIR. It was an extremely xerophytic type with a very hairy ear and similar leaves. A herbarium specimen of it exists at VIR but even that is threatened.

The extinction of durum wheat without ligula (*T. durum* convar. *aglossicon* Flaksb.) from Cyprus was prevented. In its area of origin on Cyprus this type is already extinct. In bulk populations of wheat accessions single plants of this type occurred. Such plants were singled out and received their own number in the VIR catalogue.

The Caucasus comprised more than ten botanical varieties. Of these only one or two are still part of the VIR collection. The VIR collection of Ethiopian wheats has also lost many varieties. Every botanical variety has to be preserved as a single accession, if the original landrace, which contained several varieties, is not reproduced under the conditions which are similar to its natural area of origin.

The taxonomical category 'varietas' was introduced by Fr. Alefeld and Fr. Körnicke for crop plants and is based on the differentiation by distinct characters. This category makes it possible to orientate quickly and properly in the diversity of a given area.

At the same time such classification delivers clear information for the given species with respect to the Law of Homologous Series in variation The basic taxonomical category, however, is the species. For geobotanical investigations in wild plants the use of more detailed categories, *i.e.* infraspecific taxa, is not essential.

In the early days researchers concentrated on the level of the genera; later the species level was elaborated. At present taxonomists of cultivated plants should focus on the infraspecific level.

The taxonomical classification of cultivated plants depends on the methods on which it is based, and the attention which was paid to a given species. In general the economic relevance of a species favours scientific interest.

THE CLASSIFICATION OF WHEATS

As early as 1935 Vavilov stated that "a basic handicap of all genetic investigation in wheat, as well as in other plants, is the accidental choice of the material... and the neglect of the wide range of geographical variation." For crops, which have a short history of domestication, the characterization of a cultivar can be as general as for the botanical species.

Wheat has been the basic element of food for humans for 10,000 years, and is to be found nearly all over the world. The resulting range of variation of wheat is astonishing. The formation of new infraspecific varieties is a continuous process during cultivation of the species.

The modern techniques used in plant breeding of today never led to formation of such varieties. The latest systematical overview for wheat was and it differs from the system established by Flaksberger In particular, the latest system was cleared of contradictions and inconsistencies.

The first successful approach to such a classification of wheat was done by Flaksberger in 1915. The system proposed by Percival does not differ very much from the latter. The ideas about the infraspecific differentiation of the species *T. aestivum* were very much changed owing to the expeditions of Vavilov to Central Asia, Iran, Afghanistan and India. The enormous diversity of wheat in Southwest Asia caused Vavilov to revise the value of the investigated characters of wheat for systematics. He noted that the endemic wheats are characterized by complexes of traits, which themselves are connected with distinct areas.

Table: Taxa in the Genus *Triticum*.

Species	Subsp.	Convar.	Subco-nvar.	Var.	Forms	Ecological groups
T. aestivum	2	3	4	194	15	23
T. aethiopicum	3	5	-	203		
T. araraticum	2	-	-	13		
T. boeoticum	2	-	-	61		
T. carthlicum	-	-	-	18	3	
T. compactum	2	3	4	96	2	9
T. dicoccoides	-	3	-	25		
T. dicoccon	4	4	-	64	2	
T. durum	2	6	3	120	30	12
T. ispahanicum	-	-	-	2		
T. jakubzineri	-	-	-	1		
T. karamyschevii	-	-	-	2		
T. macha	-	2	2	14		
T. monococcum	-	-	-	14	6	7
T. petropavlovskyi	-	-	-	4		
T. polonicum	2	2	-	41	1	3
T. sinskajae	-	-	-	1		
T. spelta	2	2	-	55	-	2
T. sphaerococcum	-	-	-	17		
T. timopheevii	-	-	-	4		
T. turanicum	-	-	-	20	-	2
T. turgidum	-	2	-	71	-	5
T. urartu	-	-	-	6		
T. vavilovii	-	-	-	7		
T. zhukovskyi	-	-	-	1		
Total = 25	**21**	**32**	**13**	**1054**	**59**	**63**

The endemic forms of Pamir Mountains, for example, can be distinguished by eligulatum forms and forms with more or less inflated ears. The characters themselves are connected with each other.

By studying variability in wheat, Vavilov determined the hierarchy of characters according to their taxonomic value. In the species *T. aestivum* there is a complex of characters connected with difficult threshing and stiff ears.

These traits are always accompanied by several others: rough stalks and ears, xerophytic type. Such wheat is typical of southwestern Asia. Types with easy threshing ability, on the other hand, are peculiar to Europe and areas of less continental climate of Asia (subsp. *indoeuropaeum* Vav.).

These are the two main geographical groups of wheat, which can be distinguished. Vavilov and Flaksberger regarded them as subspecies, using different names for them.

The variation led to a more detailed insight, and allowed complete description of the geographical-botanical structure by an ecogeographical system of classification. This system also uses the more detailed taxonomical unit 'varietas'. The necessity to use more detailed taxonomical units

Table: Taxa in the Genus *Triticum*.

Species	**Subsp.**	**Proles**	**Sub-proles**	**Groups**	**Greges**	**Var.**	**Forms**
T. aestivum (T. vulgare)	2	15	5	5	25	128	
T. carthlicum						10	2
T. compactum	2	3	1	1	21	68	
T. dicoccoides	2	3				25	22
T. dicoccon	5	6	3		4	65	8
T. durum	2	21	10	11	35	131	15
T. macha				2		8	
T. monococcum		3			11	2	11
T. polonicum	2				9	24	
T. spelta		2			9	3	
T. sphaerococcum						6	
T. spontaneum	2					23	46
T. timopheevii						2	
T. turgidum		2	5			19	138
Total = 14	**19**	**58**	**19**	**19**	**133**	**633**	**104**

After the analysis of the European subspecies (subsp. *aestivum*) as well as the Asian subspecies (subsp. *hadropyrum*) it became obvious that the awned varieties of *T. aestivum* mostly belong to the semi-rough-eared type of wheat. the Asian subspecies is of greater polymorphism. In particular, Middle Asia is a subarea of intense evolution of different types of *T. aestivum.*

The subspecies *aestivum* is younger. Not consisting of so many ecogeographical groups, it is, nevertheless, characterized by very contrasting ecological groups.

Length of the vegetative period, winter hardiness, resistance to diseases and other characters vary to a great extent. This subspecies covers a greater area, and stretches all over the European continent. The European subspecies has been affected by different types of ecological conditions, by different types of agriculture and by modern plant breeding. Places with a long tradition of wheat cultivation, resulting in special ecogeographical types, could be singled out.

At present these landraces are to a great extent used as basic material in plant breeding. *Triticum compactum,* having much in common with bread wheat (*T. aestivum*), was widely cultivated in the past and concurrently developed in environments similar to those of bread wheat; thus it repeats the polymorphism of bread wheat.

Reduction of the area of distribution of *T. compactum,* which took place in the remote past, and restriction of cultivation of this wheat mainly to mountainous zones led to elimination of sharply contrasting groups. No distinct isolation of subspecies is observed within this species. Flaksberger supposed that the process of differentiation in *T. compactum* had stopped in those distant historical times when cultivation of this wheat started to be replaced by more productive bread wheat. The system of infraspecific classification of *T. durum* is completely different. This wheat species is, after *T. aestivum,* the species with the widest range of geographic distribution Virtually different is the differentiation pattern of the second most important wheat species in terms of distribution, *T. durum* There is no complicated branching structure like the one observed in bread or compact wheats.

The area of durum wheat stretches along the Old Mediterranean from west to east. Durum wheat has no distinctly isolated ecological groups. The most definite are morphological differences in khoranka wheats (subsp. *horanicum*), which are characterized by a certain genetic isolation (in crosses with proper durum wheat, anomalies and sterility of the progeny may be observed). This feature preconditioned their separation into a subspecies of *T. durum,* known as subsp. *horanicum* Vav. Basically durum wheat is united by the presence of numerous common traits having similar manifestations along the whole stretched area of its distribution.

However, separate distinct features and their sets have a local character, *i.e.* they are geographically attributed: isolation of groups regulated by selection took place long ago. To describe such type of infraspecific structural subunits the taxon of 'convarietas' (group of varieties) was used. The intraspecific classification of other wheat species (*T. aethiopicum, T. turgidum* and *T. polonicum*) is less complicated and the number of botanical varieties is limited (Boxes 4 and 5). Owing to the indiscrete continuity of evolutionary processes there are almost always transitional forms (most easily preserved by humans in the case of self-pollinators).

5

Development of Sustainable Pest Management Strategies

Development of sustainable pest management strategies in agriculture has become necessary in view of increasing non-viability of chemical based approach. Among various approaches for the purpose, policy makers have paid far more attention to biotechnological alternatives. The first transgenic variety was approved for commercial trials in cotton.

However, some companies incorporate the Bt gene from the approved varieties in other varieties and released such hybrids to the farmers. It so happened that these hybrids though illegal and released unethically proved more remunerative to the farmers. As if this was not enough, farmers made crosses of this hybrid developed by Nav Bharat Seed Company and developed their own locally suited varieties.

The paper describes the ethical, business and political dimensions of agriculture biotechnology in India with specific reference to the experience of Bt cotton in Gujarat. The neglect of IPM, herbal pesticides and bio control methods becomes even less justified when state not only tolerates but also encourages widespread experimentation of Bt cotton without any regulation or monitoring. Implications for future policy for technological change.

This chapter deals with the case of Bt cotton in Gujarat, India, as an illustration of how public policy chickens out when large-scale violation of ethical and scientific norms takes place with positive business outcomes.

People have begun to take technological decisions into their own hand. Entrepreneurial spirit has created a unique case of farmer participatory research where on farm crosses are being made between Bt cotton varieties (including the ones which have been released without approval by the regulatory agencies) and other released varieties. New production cycles are being created by farmers who have extended the life of the crop from six months to nine months to reap advantage of continuous flowering and thus higher yield.

All this has happened in a completely unauthorized manner, with full public knowledge and despite complaints of Monsanto and MAYHCO about Nav Bharat

Seed Company having 'stolen' their Bt gene. Farmers are happy, politicians do not care and regulatory agencies are satisfied that sending a few committees to enquire is all that they needed to have done.

This article presents a review of important studies in the context of the Bt adventure in Gujarat and discusses the findings obtained in a farmer survey on experience and perception of Bt cotton in Gujarat.

In a developing country with overflowing food grain stocks, a liberalizing economy set on a growth path, and techno-bureaucracy willing to pursue new technologies with a reasonable open mind, the situation seems ripe for agricultural biotechnology.

The investment in this sector is picking up and the private sector has been given a pivotal role along with synergistic public investment. Top Indian policy makers argue that biotechnology will provide food security to the hungry but do not explain as to why should hunger exist at all in India with its overflowing food stocks, if the issue was only supply! Activists also make a case for biotechnology so long as it increases productivity. The case of Bt cotton in Gujarat shows that this new technology indeed increases productivity, profits of the farmer and eliminates the need for excessive sprays of chemical pesticides.

Yet, when good economic returns follow a bad ethical practice in terms of technological change (not respecting environmental regulation and monitoring, nor intellectual property rights), do ends justify the means? The answer begs for possible alternatives to Bt cotton:

- *The agro-chemical industry and conventional agriculture*: The chemical pesticide industry makes its living so to say, by heavy inputs in cotton (reportedly 40-50 per cent of all the pesticides in India is consumed by hybrid cotton alone). It does not even educate workers and farmers about (a) safe ways of spraying (not almost a single hoarding in the entire country on the subject though there are thousands of bill boards advertising the use of the chemical pesticides), (b) disposing pesticides containers (except in small print in the publications accompanying the pesticide bottles), and (c) practicing integrated pest management –IPM-practices. Ethical violations in chemical pesticide industry are more conspicuous, have hazardous results and are more certain than other new technologies-in fact consequences of excessive use of chemical pesticides on human and environmental conditions are well documented as against the environmental consequences of biotechnology which are still being probed Yet, similar public debate does not ensue.
- *Integrated Pest Management* (IPM): Despite having achieved excellent results through publicly funded IPM, the investment on IPM in Gujarat remained static for the last three years-around 75000 US$

per annum that is less than one percent of the total agricultural budget in Gujarat. This is very likely to be the case at the countrywide level as ascertained in informal dialogues with senior policy makers in the Ministry of Agriculture. IPM offers an alternative to biotechnological solutions in pest control and a cheaper alternative at that. But public policy support for IPM is quite weak.

Even though the adoption of Bt cotton in Gujarat has benefited farmers significantly, the public policy neglect of safer, cheaper and more affordable technological alternatives has obvious ethical implications too.

The Department of Biotechnology (DBT) of the Government of India is aware of these alternatives and has made hardly any investment in making even fair comparative trials of Bt cotton with other competing technologies.

Not to mention the almost total absence of investment in public education about the likely hazards of taking up such large scale un-guarded, un-monitored and un-evaluated trials by farmers of the Bt crosses with local hybrids.

Undoubtedly, the experience of Gujarat will be recalled in the history of biotechnology as one of the largest trial (with full public knowledge and without any responsible monitoring or evaluation by public agencies at similar scale) of an illegally released technology ever done by people themselves, oblivious of any environmental or other consequence.

OVERVIEW

For example, Cornell University researchers are developing a rice variety that yields 20 percent to 40 percent more food per acre and a tomato that produces 48 percent higher yields per acre.

Biotechnology promises to play a critical role in reducing blindness caused by vitamin A deficiencies. Every year in Southeast Asia, one quarter of a million children go blind because of a lack of vitamin A in their diets.

Worldwide, 124 million children suffer from vitamin A deficiency. Increasing the intake of vitamin A could prevent one to two million deaths per year and eliminate numerous health problems.

Poverty and lack of infrastructure make the traditional methods of vitamin consumption (a varied diet or supplements) impossible. Agricultural biotechnology, however, can help to address the problem of childhood blindness and death by ensuring that people get this nutrient in the key staple of their diet, rice.

Researchers have spliced into the genome of rice, the beta-carotene gene from a daffodil to create “golden rice” — rice high in beta-carotene, which the human body converts into vitamin A. This is among the most well-known examples of biotechnology’s benefits. Countless other genetically engineered crop varieties are being developed to address similar problems of poor nutrition or productivity.

HOW AGRICULTURAL BIOTECHNOLOGY IMPROVES PLANT SAFETY

Many biotechnology skeptics contend that such engineering is unsafe because it is "unnatural." Ironically, biotechnology is actually an improved method for plant breeding. Most of what we eat today has already undergone gene manipulation under the traditional breeding methods.

The traditional methods and biotechnology differ in that traditional methods are cruder because they do not give researchers control over all the genes they crossbreed. Some desirable characteristics might result with such methods, but undesirable ones might as well. It could take a long time of continual breeding to eliminate the most problematic traits and focus on the best ones.

Compared to these more "traditional" methods of hybridization, genetic engineering is a wonder of precision and accuracy. Genetic engineering enables scientists to pinpoint the genes that express a desirable characteristic in one plant and to insert those genes into the genome of another plant. Scientists mark the insertion point so that they know where these novel genes are located in the genome of the new plant. While there are several different methods of genetically engineering plants,

GENERAL BIOTECHNOLOGY INDICATORS

Since 1999, organizations such as the Organization for Economic Co-operation and Development (OECD) and the Organization of American States (OAS) began to compile statistics to describe the situation of the countries that are developing biotechnologies.

A general view shows that countries in Latin America and the Caribbean (LAC) have significant gaps with countries of the North, especially in patterns of patenting and licensing technologies, and that the transferal of biotechnologies is of vital importance, especially those biotechnologies developed for the agricultural sector.

In LAC countries, measurements mainly include the analysis of the capacity of groups and research centers, their working fields, the number of enterprises and the market perspectives, whereas the statistics of OECD measure mainly the private investment, the generation of investor's capital in terms of new companies, money and employment, and the obtainment of patents among other indicators, which are not so representative for us due to the differences in economies and development systems.

An indicator to analyze the state of biotechnology in the countries is a function of the investment made for its developments and products. The United States of America has a much higher investment tendency in relation to other countries. In 1997, the industry invested around US$ 7.300 millions, being 3.7% of the total investment for research and development (NSF, 2004), and as for

2001 it reported that investment of industry in research and development of biotechnology reached US$ 16.400 millions, representing 10% of the total investments in science and technology. European countries such as Germany, with more than US$ 1.000 millions (6.7% of total), United Kingdom with US$ 705 millions (7.8% of total) and France with US$ 560 millions having a 4.4% of the total of governmental investment in R&D, follow the list with the largest assigned budget for biotechnology according to statistics of OECD.

The situation in countries members of the OECD reveals a striking difference between USA and EU. According to data provided in 2001 by the consultant company Ernst and Young, USA investment in R&D was □ 11.400 million (2.7% of DGP) whilst EU investment was □ 4.977 million (1,8% of DGP). There are 1,262 companies in USA totalling 162,000 employees and 6.7% of them are involved in research, in contrast to 1,570 companies in EU with 61,000 employees, 2.5% of them conducting research.

In the compendium of biotechnology statistics of OECD, differences can also be observed. The first one is the amount of investment in risk capitals in biotechnology, which goes up to US$ 1.182 millions in the USA, and US$ 687 millions for the EU during 1999.

The bioengineering products reached that year up to 1% of the total exports of USA with a total amount of US$ 1.340 millions. Its import of products from OECD countries, especially from Belgium, France, Switzerland and The Netherlands were that same year of US$ 970 millions. Its most important commercial partners where most of the exports are sent to include: Belgium, Japan, Canada and Germany.

This indicator allows analyzing the degree of technological delay resulting from the efforts of countries such as Colombia, where it represents approximately US$ 1.8 millions.

Exchange rate 1 US$ = $ 2.200 Col) and 3.7% of the total governmental investment of the National Biotechnology Programme (NBP) made through Colciencias during 1997, and during the following years the tendency remained negative, reaching in 1999 almost US$ 780.000 and 4.8% (OCyT, 2004). This situation somehow explains why Colombia reports 8 patents granted to research groups associated with the NBP (OCyT, 2004), whereas the USA already had more than 3.600 patents and Canada more than 500 during the year 2000 (Van Beuzekom, 2001). It could be said that the relation between the investment in biotechnology and the number of patents is more than US$ 2 million per patent in the USA, US$ 640.000 in Canada and US$ 22.000 in Colombia.

On the other hand, the development of biotechnology in LAC has been characterized by being a repetitive model where the research projects do not correspond to the real needs of production and food security, and has a low level of innovation and scientific creativity. Only 2% out of 30 scientific articles of high impact on biotechnology came from laboratories based within the region, despite LAC possessing important biodiversity resources to develop innovative products.

AGRICULTURAL BIOTECHNOLOGY REGULATION

For traditionally bred plants, regulators rely on plant breeders to conduct appropriate safety testing and to be the first line of defence against genetic alterations that might prove dangerous.

The same should be true for agricultural products developed using biotechnology. There is no reason for regulators to treat biotech plants any differently than traditionally bred plants, particularly given the fact that biotechnology provides greater control over gene manipulation. Despite fears about gene manipulation, traditional crossbreeding has altered plant genes and improved the human diet for the past 30 years.

In fact, people worldwide safely consume these plants every day. Examples include wheat, corn, potatoes, tomatoes, and countless other staples of the American diet. In 1984, the biotechnology industry began experimenting with "genespliced plants" — the more advanced approach to gene manipulation that we now call "biotechnology."

REGULATORY SCHEME

At the time, the White House Office of Science and Technology Policy began crafting a frame-work wherein existing federal agencies would regulate genetically engineered organisms on the basis of their characteristics, not the method of production — thus wisely deferring to the scientific community's judgement that regulation ought to address the products, not the process of biotechnology. Engineered organisms would not require extra scrutiny simply because genetic engineering produced them (the process).

Instead, they would be subject to heightened scrutiny only if the individual organisms expressed characteristics that posed some conceptually heightened risk (the product).

The federal government divided regulatory jurisdiction among agencies already involved in agricultural, food, and environmental regulation. These include the United States Department of Agriculture (USDA), the Environmental Protection Agency (EPA), and the Food and Drug Administration (FDA).

While each of these agencies considers the characteristics of individual products in their regulation, only FDA followed the general scientific thinking that genetically engineered and non-genetically engineered products should be regulated similarly. Both USDA and EPA automatically subject all genetically engineered plants as a class to premarket approval requirements not ordinarily applied to conventionally bred plants.

DEPARTMENT OF AGRICULTURE

USDA regulates the release of all genetically engineered agricultural plants under statutes giving USDA's Animal and Plant Health Inspection Service (APHIS) the authority to regulate plants that may be or may become weeds or other nuisances — what the statutes call "plant pests."

Although the rules apply in a general sense to novel or exotic varieties of both genespliced and conventional plants, APHIS typically only requires field testing of conventional plants that are new to a particular U.S. ecosystem (transplanted from another continent, for example).

However, all genetically engineered agricultural plants face a higher regulatory bar — making it more difficult to expand the use of biotechnology. Genetically engineered crops must be field tested under APHIS regulations prior to commercialization.

For example, a new variety of corn produced with conventional hybridization requires no government-mandated field testing, but all new varieties of genetically engineered corn do, even though there is no logical reason for the regulatory disparity.

For most genetically engineered plants, APHIS requires the company producing the plants to submit notice detailing the gene or genes that have been inserted, where the plant will be tested, and other relevant characteristics of the plant prior to receiving permission to conduct the field trials.

Once the company completes field testing, APHIS reviews the results and makes a determination on whether or not the product should be "deregulated" and can be released into the market.

ENVIRONMENTAL PROTECTION AGENCY

EPA regulates plants that are modified to produce substances that act like pesticides: that is, substances used by a plant to protect itself from pests, such as insects, viruses, and fungi.

EPA's proposed rule for these "plant-incorporated protectants" is not yet finalized. In the interim, however, EPA has regulated such plants by applying its proposed guidelines, which are functionally similar to rules for the registration of synthetic chemical pesticides. Again, biotech products face higher regulatory hurdles, even though plant-incorporated protectants developed through conventional breeding are exempt from these requirements.

FOOD AND DRUG ADMINISTRATION

FDA is responsible for ensuring that food items, including foods derived from genetically modified plants, are safe to eat. Following the general regulatory framework that emphasizes product regulation rather than process regulation, FDA rightly does not treat foods derived from genetically engineered plants to be inherently unsafe. Food producers are not required to seek premarket approval from FDA unless there is a substantive reason to believe that the novel trait (or traits) in the food pose a safety question.

The initial determination of safety is left to the producer, but FDA has encouraged producers to consult with agency scientists prior to marketing a food produced with biotechnology to ensure that the appropriate determination is made. Recently, FDA published a proposed rule that would require producers to notify the agency at least 120 days before marketing, which may be a signal that FDA could abandon its more reasonable approach in the future.

In addition, FDA does not require labeling of foods derived from biotechnology unless the genetic insertions so alter the food that the common name no longer describes it adequately. Examples of this phenomenon would include such alterations of the food product that would raise a safety issue or change the product's nutritional composition or its storage or preparation characteristics.

LABELING ISSUES

Some activists, however, argue that the government should mandate the labeling of all genetically engineered foods. They assert that consumers have a "right to know" how their foods have been altered, and that a mandatory label would best allow consumers to choose between genetically engineered and conventional foods.

Biotechnology advocates have argued against mandatory labeling because such requirements raise food costs — something that mostly harms lower-income Americans and people on fixed budgets.

Perhaps more important, while biotech products are not substantially different from other products, special labels would likely make consumers think these products were more dangerous.

Hence, rather than serving educational or "right to know" purposes, such labels promise to simply confuse consumers.

A government-mandated label on all genetically engineered foods also would raise important First Amendment free speech issues. In 1996, the U.S. Court of Appeals, in *International Dairy Foods Association et al. v. Amestoy*, ruled unconstitutional a Vermont statute requiring the labeling of dairy products derived from cows treated with a genetically engineered growth hormone, noting that food labeling cannot be mandated simply because some people would like to have the information.

"Absent ... some indication that this information bears on a reasonable concern for human health or safety or some other sufficiently substantial governmental concern, the manufacturers cannot be compelled to disclose it."

In other words, to be constitutional, labeling mandates must be based in science and confined to requiring disclosure of information that is relevant to health or nutrition. Furthermore, consumers need not rely on mandatory labeling of biotech foods to truly have a choice. Realworld examples show that market forces are fully capable of supplying information about process attributes (including kosher and organic production standards) that consumers truly demand.

The same can be said about nonbiotech foods, and the FDA recently published proposed guidelines to assist producers in voluntarily labeling both genetically engineered and nongenetically engineered foods. Additionally, the USDA's newly published rule for organic certification necessarily excludes biotech products from organic food production.

Consequently, consumers wishing to purchase nonbiotech foods need only look for certified organic products.

INTERNATIONAL TRADE

While U.S. consumers do not appear to be strongly opposed to biotech foods (they, in fact, seem rather indifferent), a strong anti-biotechnology movement has arisen in several European countries over the last few years. The European Union (EU) has established strong restrictions on the commercial planting of genetically engineered plants, and European food processors and retailers are reluctant to import harvested agricultural products derived from biotechnology — thus jeopardizing the marketability of U.S. commodity grain exports.

And the EU is now negotiating for strong restrictions on agricultural biotech products in important international agreements governing food safety and environmental protection. Very strong restrictions were included in the Biosafety Protocol, finalized in January 2000.

Perhaps, more important, though, are the Codex Alimentarius Commission standards for food safety. Without convincing scientific evidence that genetically engineered crop plants pose a heightened environmental or human health risk, restrictions on agricultural biotech imports could be challenged under the General Agreement on Tariffs and Trade and the World Trade Organization.

Thus the EU has resorted to justifying its decisions on the basis of a risk-management philosophy known as the "precautionary principle," and the EU advocates inclusion of the precautionary principle in international environmental and food safety agreements (such as the Codex Alimentarius), as well as within GATT itself. Its inclusion would give nations greater latitude in restricting imports of U.S. agricultural products.

No single definition exists for the precautionary principle, but its general meaning is that when an activity raises threats of harm to human health or the environment, regulatory measures should be taken to prevent or restrict the activity even if the risks have not been demonstrated scientifically.

Although the EU asserts that the precautionary principle is an unbiased risk-management philosophy, critics have noted that its lack of definition and evidentiary standards makes it all too easy to abuse for the purpose of masking trade protectionism, and that its very approach to risk management is inherently flawed and may, in fact, increase net risk. Nevertheless, the precautionary principle has already been included in several international environmental treaties, including the Convention on Biological Diversity and the Biosafety Protocol.

Even as underdeveloped nations clamor for biotechnology applications, and as countries like China continue to experiment with and use agricultural biotechnology, opponents of agricultural biotechnology in the West, particularly Europe, attack it as an unnatural process that will destroy the world, not better it. They argue that biotechnology should be heavily regulated, if not banned. Already, however, genetically engineered plants are subject to strict regulatory oversight that is equal to or greater than that advocated by the vast majority of scientific specialists.

Additional regulation will slow down research and development of genetically engineered crops, keep beneficial products off the market, and raise the cost of products that do make it to consumers. Furthermore, the inclusion of similar restrictions — or inclusion of the precautionary principle — in international agreements will greatly impact the international trade of agricultural goods and delay their introduction into the marketplace. Each of these problems could prevent this technology's benefits from being introduced to industrialized nations and, more important, the developing world.

TWENTY YEARS OF MODERN AGRICULTURAL BIOTECHNOLOGY

In 1973, Cohen and Boyer transferred a gene from one organism into another. In 1982, the first biotech plant, an antibiotic resistant tobacco, was developed.

In January 1983, at a meeting of genetic researchers in Miami, three different teams reported success in using Agrobacterium tumefaciens, a bacterium, to carry new genes into plant cells, heralding the dawn of modern agricultural biotechnology.

Agrobacterium tumefaciens, described as a "natural genetic engineer", splices its own genes into host plant cells. This pathogenic bacterium was now converted into a pack mule, to carry new, foreign genes into plant cells, minus the disease and this became the most common means of producing Genetically Engineered Organisms (GEOs).

Field tests for GE crops resistant to pests and pathogens were first conducted in the US, in 1985. Accoordinated framework for the regulation of GEOs was established and the first GE tobacco was released, in 1986. The US Department of Agriculture published guidelines for field trials of GE crops in 1991. On approval from the Food and Drug Administration, Flavour Saver, the first GE tomato, with a longer shelf life, was on the US markets in 1994. During 1995-96, GE soybean, corn and cotton were approved for commercialization, in the US.

A number of GE crops, developed for pest, pathogen and herbicide resistance, are now commercially cultivated in several countries. Rice with pro-vitamin A, higher iron content, or human milk proteins and potatoes with high protein content are in various stages of development. Tobacco plants producing functional human haemoglobin and bacteria that produce human insulin have been developed, as well as plants with vaccines against rabies and other viral diseases.

Food grain crops that withstand drought and salinity are high priority research, and so are those for high yield. A GE tobacco plant detoxifies soils contaminated by explosive residues, providing solution to a frustrating environmental problem in countries ravaged by armed conflict.

Now more than 70 biotech agricultural crops that have been approved for use in North America, including varieties of soybeans, cotton, canola, corn, potatoes, squash, tomatoes and papaya. About six million farmers in some 17 countries now cultivate GE crop on about 125 million acres, a 30 fold increase over 1996. By end of the year 2002, six GE crops planted in the US (soybeans, corn, cotton, papaya, squash and canola) produced an additional four billion pounds of food and fibre on the same acreage, improved farm income by US $ 1.5 billion and reduced pesticide use by 46 million pounds. In 2003 in the US, 80 per cent of soybean acres will be planted with biotech varieties.

A number of activist groups vehemently attack GE technology and its products on grounds of safety to humans and the environment, and costs of technology transfer and its reach to the needy. Products of agricultural biotechnology bear the brunt of this ill-informed, unscientific and prejudiced onslaught much more than GE products related to health or industry.

In 2001, the European Community released results of a 15-year study, costing US $ 64 million, and involving more than 400 research teams and 81 projects. This report concluded that GE products pose no more risk to human health or the environment than conventional crops. So far, extensive and intensive research on the probable risks of GE technology has not brought out any adverse effects and none of the fears expressed by anti-tech activists were proved even marginally. Nevertheless, caution and examining issues case by case, is the watch word of technologists, who are aware of their responsibilities.

For a number of products, technology transfer is free of costs for developing countries, as for example Golden Rice, the rice with pro-vitamin A. It is the responsibility of the Governments of the respective countries to bear the costs of developing local varieties and to reach the products to the needy at an affordable cost. In India, three GE varieties of pest resistant cotton were approved for commercialization, a year ago. Approval for other GE varieties of cotton and GE mustard was deferred twice by the Genetic Engineering Approval Committee (GEAC), the highest authority on the issue.

In India, the level of public awareness of the realizable benefits and probable risks of GE products is abysmally low. The functioning of the GEAC leaves much to be desired. Taking advantage of this hazy situation, mixing up economic, social and political issues with science, and even using such vagaries of nature as the severe drought, several groups of vested interest have created mistrust, confusion and scare. Trashing a very promising technology this way does not augur well for the future of Indian agriculture. This results only in denying the benefits of technology to the farmers and consumers.

The next 20 years of plant biotechnology is expected to improve quality of life, through improved foods, pharmaceuticals, new industrial materials and a better environment.

These benefits should reach the people of the developing world, who need them most. For India to benefit from these developments, it is urgent and essential that the public is educated on the realizable benefits and probable risks related to GE products. We also need to reorganize the mechanism of regulating GE products by providing for an expedient, transparent and responsible authority. The media have a very important role to play in this regard. Only an informed and reassured public can make viable choices. Every effort should be made to provide such an opportunity to the consumer.

MANAGING AGRICULTURAL BIOTECHNOLOGY IN COLOMBIA

The international scenario for biotechnology shows a rapid tendency at industrialized countries in the increase of publications, patents, enterprises and novel solutions for the industry, the environment, health and agriculture. Nevertheless, Colombia has an important delay in relation to the international scientific development and the capacity to generate wealth and services for its productive systems.

This delay has been one of the concerns of the country's policy during the last years, and more precisely since 2002, when for the first time biotechnology was included in a National Development Plan as one of the mechanisms for competitiveness and the use of biodiversity and genetic resources.

This chapter is the result of a survey conducted in 2005 aimed to provide an overview of agrobiotechnology in Colombia to be included in the compendium of case studies organized by the FAO's Regional Office for Latin America and the Caribbean (LAC) and the Network for Technical Cooperation in Agricultural Biotechnology in Latin America and the Caribbean (REDBIO/FAO).

The command of skills on classic biotechnologies in Colombia has allowed the strengthening of capacities, especially with micropropagation and cell and tissue culture. Nevertheless, the transition towards modern biotechnologies has been difficult, especially due to the scarcity of financial resources and the access to new technologies and specialized knowledge. Neither has the country a critical mass of economists, managers and lawyers to integrally support the processes of technological supervision, competitive intelligence, market access, national regulations and international treaties on intellectual property rights and biosafety.

The attention that necessities of small producers deserve, together with the improvement of the intellectual property rights, public perception and biosafety framework on genetically modified organisms (GMOs) for human and animal consumption, are some of the fields on which we should be working in the country through international networks, with an aim to reach the benefits that use of modern biotechnology brings along, as some researchers in this field have highlighted. In this way, the current technological gap existing between countries of the South and the North could be bridged.

THE STATE OF BIOTECHNOLOGY IN LATIN AMERICA AND THE CARIBBEAN

Bourinbaiar (2006) recently reviewed developments achieved outside EU that were presented at the BIO 2005 meeting and highlighted progresses done in Argentina and Chile among LAC countries. A comprehensive review was undertaken based on five important books published between 2003 and 2005 by multilateral organizations, showing comparisons and experiences among several LAC countries in terms of scientific and technological capacity, political situation, biosafety, commerce, biodiversity and international markets.

The first study, made by CamBioTec in 14 Latin American countries, Argentina, Bolivia, Brazil, Chile, Colombia, Costa Rica, Cuba, Ecuador, Guatemala, Mexico, Peru, Paraguay, Uruguay and Venezuela, clearly shows that for 2002 the biotechnological industry was represented by 432 companies (including less than 20 companies in the agricultural sector in countries of the Andean Community of Nations (CAN)), where the most outstanding are the BioMinas cluster in Brazil, the agro biotechnological pole of the National Institute of Agricultural Technology (INTA) from Argentina and the Western Biotechnological Pole in Cuba. The companies mostly involved with the agricultural sector are in the fields of cell biology and microbiology, with emphasis on the propagation of plant materials to lower cost and time production. Regarding national regulations for the protection of intellectual property rights, microorganisms can be patented in Brazil and Mexico, while in Chile it is possible to patent biotechnological processes.

From the second study, financed by Iberoamerican Programme for the Development of Science and Technology (CYTED) during 2003, it can be concluded that while the majority of researchers work in the agricultural sector, innovations are not so representative when compared to the pharmaceutical sector. It also shows that one of the topics where the policies of the countries (with the exception of Brazil) have had less result is in the creation of new enterprises with a biotechnological base. This is due to the shortage of monetary resources and the lack of continuity in the national programmes and plans (REVYDET, 2003). The third one, published in 2003 by OAS, CamBioTec, the USA Embassy in Chile and the Chilean Government, shows the results on different topics related to the application of the Cartagena Protocol and foods derived from genetically modified plants. The study shows experiences that allow us to have an overview of the situation, the perspectives and challenges that modern biotechnology brings to agriculture and it pretends to facilitate the discussion on the risks and benefits of biotechnology for the region.

The fourth study was financed by OAS and published by the Colombian Observatory of Science and Technology (OCyT), presents statistics from Colombia, Costa Rica, Mexico and Venezuela together with the comments from different experts on the construction of indicators in biotechnology. In the different comparisons made with individuals dedicated to R&D, research facilities, training programmes, projects and products, there is an evident lack of univocal usage of the notion of biotechnology, while in some countries activities of second generation are accounted, and in other countries only the projects involving the use molecular biology and genetic engineering are accounted.

Finally, during 2005 the Andean Development Corporation (CAF) produced an analysis of the markets and capacities of the five countries of the CAN, to make use of biodiversity. From the CAF's study it can be highlighted that from the 567 research groups from the five participant countries, half of them work for the agricultural sector. According to the indicators developed by Dr. William Roca for the CAF, the Andean region has reached a 52% of development of the scientific and technological capacity considered optimal to make use of biotechnology and biodiversity. Regarding the financial tendencies, the study reports that from the 22 groups selected from the five countries that get yearly funds for research, they get on average US$ 97.500, within a range from US$ 30.000 to US$ 1.5 millions (CAF, 2005)

In the Network for Technical Cooperation in Agricultural Biotechnology in Latin America and the Caribbean (REDBIO/FAO) there are four case studies at the national level for Argentina, Bolivia, Ecuador and Peru available at www.redbio.org/estud_casos.htm. According to these studies, we can see in the case of Argentina that modern biotechnology has had an enormous influence from multinational companies controlling the market of transgenic seeds, and that there has been important progress on surface area cultivated with GMOs. Bolivia shows very little progress in biotechnology, especially on laboratories involved in research, enterprises involved in agricultural produce marketing and as licenses for access to genetic resources. Ecuador, on the other side, has had some progress in the research on Andean roots and tubers and on tree tomato but nevertheless, its progress have been very limited, especially with the introduction of GMOs and the use of biodiversity and genetic resources. Finally Peru has only one request for introducing a GMO from the International Potato Centre (CIP), and its progress at the institutional and political levels has not allowed yet the strengthening of the capacities needed for the effective use of the potential represented in biotechnology. About the current capacity of the LAC, the data bank CATBIO from the Network for Technical Cooperation in Agricultural Biotechnology in Latin America and the Caribbean (REDBIO/FAO), founded in 1991, provides information on 738 laboratories located in governmental, private and academic institutions from 32 countries in the region, and from 4300 professionals working in areas related to research, regulation, extension, education and diffusion of biotechnology, all of them integrated in the module *profREDBIO*.

AGRICULTURAL BIOTECHNOLOGY: UPDATED BENEFIT ESTIMATES

U.S. farmers have adopted genetically modified crop varieties rapidly since their introduction. By 2000, roughly one fifth of U.S. corn acreage, over half of the soybean acreage, and almost three-quarters of the cotton acreage was planted to crops genetically modified to be resistant to insects and/or herbicides.

This report updates our previous estimates of the benefits associated with the adoption of genetically modified crop varieties. Updated impact estimates have been calculated for corn, cotton and soybeans. A discussion of potential impacts for potatoes is included. Each of these crops delivers a unique set of benefits to growers who adopt them. These benefits largely depend on pest control issues particular to each crop and whether other effective and affordable pest control options are available. Understanding the reasons why growers are adopting these varieties is critical in evaluating the their impact.

Insect resistant Bt corn varieties have allowed farmers to control the European Corn Borer, an insect pest that is difficult to control using conventional insecticides because it tunnels into the corn stalk soon after hatching, and is protected from the insecticides. Prior to the introduction of Bt corn, few growers were spraying for the corn borer.

Instead, yield losses sometimes reached 300 million bushels of corn per year. With Bt corn, losses to the corn borer are eliminated. The primary benefit of Bt corn varieties has been increased yields. In 1999, it is estimated that 66 million bushels of corn were saved from the corn borer, or the equivalent of production on nearly 500,000 acres. Modest reductions in corn insecticide use have also been observed since the introduction of Bt corn.

Cotton growers have adopted genetically modified varieties faster than growers of any other crop. Both insect and herbicide resistant varieties have been adopted widely. Insect protected varieties of Bt cotton provide control of three of the most destructive insect pests in cotton: tobacco budworm, cotton bollworm and pink bollworm. Growers were managing these pests with several insecticide treatments per year. In some areas, the efficacy of these insecticides was diminished by the development of insecticide resistant insect populations. Bt cotton has allowed growers to reduce insecticide use and attain better control of these pests, which has resulted in increased yields. It is estimated that cotton growers reduced insecticide use by 2.7 million lbs and made 15 million fewer insecticide applications per year due to the introduction of Bt cotton. Cotton production has also increased, by 260 million lbs per year. Net revenues are estimated to have increased by $99 million in 1999.

Herbicide resistant cotton varieties provide growers with effective weed control programmes that eliminate some of the problems associated with conventional programmes. Until 1995, cotton growers did not have any broadleaf herbicides that could be used over a growing cotton crop without causing crop injury. Instead, herbicide treatments on small cotton required time consuming treatments using special equipment to treat weeds between the rows without contacting the crop. With the introduction of herbicide resistant cotton, growers can use broad spectrum herbicides over the growing crop with minimal crop injury. The introduction of herbicide resistant cotton varieties has led to a reduction in the number of herbicide applications made by cotton growers, by 19 million in 2000.

Insect and virus resistant potato varieties have been introduced that have great potential to decrease insecticide use in that crop. However, the recent introduction of a highly effective conventional insecticide and the refusal of processors to accept genetically modified potatoes have limited the adoption of these new varieties. Finally, herbicide tolerant soybeans offer growers effective weed control with a simple, flexible programme that has allowed many to reduce weed control costs. Prior to the introduction of herbicide tolerant soybeans, growers chose from many herbicides, often applying three or more active ingredients, some of which would cause damage to the growing soybean plants, or cause harm to corn crops that commonly follow soybeans. Herbicide tolerant soybeans allow growers to rely on one herbicide to control a broad spectrum of weeds without harming the current or rotation crops. The primary benefit of herbicide tolerant soybean varieties has been a reduction in weed control costs, of $216 million per year in 1999. Growers have also reduced the number of herbicide applications, by 19 million in 1999.

U.S. farmers have adopted genetically modified crop varieties rapidly since their introduction in the mid-1990's. Genetically modified varieties of three major crops including corn, cotton and soybeans have been planted in the United States. A genetically modified potato variety has also been planted on limited acreage. Each of these crops delivers a unique set of benefits to growers who adopt them. These benefits largely depend on pest control issues particular to each crop and whether other effective and affordable pest control options are available. Understanding the reasons why growers are adopting these varieties is critical in evaluating the impact these technologies have had on US agriculture. Previous estimates of aggregate benefits to U.S. farmers indicate higher yields, lower costs and ease of management. In addition, pesticide use has declined since the introduction of genetically modified varieties. This report updates previous estimates of aggregate benefits resulting from farmers' adoption of genetically modified varieties for corn, cotton, potatoes and soybeans. The aggregate impacts of the introduction of genetically modified crop varieties. These tables present aggregate estimates of changes in net revenue, increases in production, decreases in pesticide use and decreases in pesticide acre treatments.

An application acre is the number of different active ingredients applied per acre times the number of repeat applications, and differs from the number of trips over the field, as one trip across the field to apply two active ingredients is treated as two applications, as is two treatments each containing a single ingredient. Bt corn insecticide reductions in number of acres treated HT cotton-herbicide tolerant cotton, including BXN and Roundup Ready varieties.

CORN

Bt field corn varieties were introduced for planting by U.S. farmers in 1996, offering built-in insect control. Adoption of Bt corn varieties was swift, rising to 26% of corn acres planted by 1999, the fourth year on the market. Adoption decreased to 19% in 2000, which is believed to be due largely to historically low target pest pressure in 1998 and 1999. The adoption of Bt corn since its introduction in 1996. Corn is the largest acreage crop grown in the U.S. Planted acreage totaled 80 million in 2000, which represents approximately one-quarter of the acreage planted to all crops in the U.S. Average corn yields totaled 134 bushels per acre in 1998 and 1999 (USDA NASS 2000a). Corn for grain production was estimated at 9 billion bushels in 1999 (USDA NASS 2000c) with a total value of production of $18 billion. This represents approximately 20% of the value of all crops grown in the U.S. (USDA NASS 2000d). All 48 coterminous states have corn acreage, and, in many states, corn is the single most important crop in terms of acreage and production value. Corn production is concentrated in the Midwest, where ten states account for 85% of U.S. acreage and production. Individually, the states of Illinois and Iowa account for over 10 million acres of corn each (USDA NASS 2000a).

The major use of corn produced in the U.S. is as a livestock and poultry feed, accounting for 60% of use, while food, seed and industrial uses (including sweeteners, fuel alcohols and starch) account for approximately 19% of use. Exports account for the remaining 21% of use (USDA NASS 2000b). The U.S. is by far the largest corn producing country, growing 40% of the world's total production. China is the second largest producer, growing 21% of the world's total. Brazil, Mexico, France and Argentina are also major corn producing countries. The U.S. dominates the corn export market, accounting for 75% of the world's total exports. Major export markets for U.S.-grown corn are Japan (29%), Korea (12%), Mexico (10%) and Taiwan (8%). Western Europe accounts for less than 1% of U.S. corn exports.

Bt corn varieties were genetically modified to express an insecticidal protein from the soil bacterium *Bacillus thuringiensis* (Bt). The Bt protein is selectively active against lepidopteran, or caterpillar insects, including the European Corn Borer (ECB), a major insect pest of field corn in the U.S. The ECB ingests the Bt protein when it starts to feed on the plant. The toxin binds to the gut membranes, which causes the ECB larvae to die. Due to the difficulty in scouting for this pest and the importance of timing insecticide application before the caterpillar bores into the corn stalk and is protected from insecticides, it is estimated that less than 5 percent of field corn acreage in the U.S. "Corn Belt" was being treated with insecticides for the ECB prior to the introduction of Bt corn varieties.

Researchers have sought alternative methods to control the corn borer, although none has proven effective on a wide scale. Traditional breeding efforts to select for natural resistance to the corn borer resulted in the development of varieties with intermediate levels of resistance that were widely used into the mid 1970s. However, acreage planted to these varieties decreased dramatically due to the introduction of much higher-yielding susceptible hybrids, which could produce higher yields than the resistant hybrids, even after sustaining high levels of damage from the corn borer. Another extensive research programme was undertaken by the U.S. Department of Agriculture (USDA) to identify natural predators of the corn borer. The programme resulted in the introduction of 24 species of parasites in U.S. corn production areas, of which only six became established. While these beneficial insects have provided control of the corn borer in some years in some areas, their impact has been limited.

Several different types of Bt field corn have been registered. The U.S. Environmental Protection Agency (EPA) has registered five unique Bt corn products: 176 (Novartis Seeds and Mycogen Seeds), Bt11 (Northrup King/ Novartis Seeds), Mon 810 (Monsanto), DBT418 (Dekalb Genetics), and CBH 351 (Aventis, was AgrEvo). The Mon 810 and Bt11 events are used in production of YieldGard corn. Event 176 is sold as KnockOut by Novartis and Nature Gard by Mycogen. The DBT418 event is sold as BT-Xtra. The CBH351 event is sold as Starlink by Aventis. These events vary in the specific Bt protein that is produced, how much of the protein is produced, and where it is produced in the plant, thus affecting ECB control. The Starlink corn hybrid contains the Cry9C protein while all the other Bt hybrids contain the Cry1A(b) or Cry1A(c) protein. The Cry9C protein binds to a different site in the insect's gut. Cry9C is also toxic to black cutworms. The registration for Starlink varieties was voluntarily withdrawn by Aventis in October 2000. Starlink varieties had been planted on less than 1% of total U.S. corn acreage.

The impacts of Bt corn include increased yields and reduced pesticide use. However, the main advantage of Bt field corn has been increased yields, as only a small proportion of U.S. field corn acreage was sprayed for the ECB prior to the introduction of Bt varieties. Yield losses due to the corn borer vary from year to year with infestation levels, which are generally unpredictable from one year to the next. Largely uncontrolled until the introduction of Bt corn varieties, the ECB has caused production losses that have ranged from 33 million bushels to over 300 million bushels per year (USDA APHIS). Bt corn varieties have been shown to provide a very high level of protection from the corn borer. Plants sustain only minute damage as the corn borer larvae attempt to feed.

The benefits that growers realize from planting Bt corn varieties depend on the level of infestation. In light infestation years the benefits may not be great, while in heavy infestation years growers will realize substantial yield increases. An average of available research results comparing yields from Bt and non-Bt corn fields indicates that growers experienced a yield advantage of approximately 12 bushels an acre in 1997, 4.2 bushels an acre in 1998 and 3.3 bushels an acre in 1999. The seed price premium, or technology fee, for Bt corn was approximately $10 an acre in 1997 and 1998 and $8 in 1999. Assuming corn prices of $2.43 a bushel in 1997, $1.95 a bushel in 1998 and $1.90 a bushel in 1999, the average income changes for Bt corn were an increase of $18 an acre in 1997, a decrease of $1.81 an acre in 1998 and a decrease of $1.73 an acre in 1999. It is expected that in 10 of the 13 years from 1986 to 1998, corn borer infestations in the Corn Belt were such that corn growers would have realized a gain from planting Bt corn (Monsanto). The aggregate cost and benefits estimates of Bt corn for 1997 through 1999.

The greatest impact of Bt corn varieties is the increase in production it provides through reduced yield losses due to the ECB. The increase in production due to reduced yield losses due to Bt corn for 1997 through 1999. In 1999, growers were able to reduce their yield losses by 66 million bushels, the equivalent to production on nearly 500,000 acres of corn production that would otherwise have been destroyed by the corn borer.

Reductions in insecticide use are also expected due to the introduction of Bt corn varieties, though these reductions are anticipated to be modest due to low levels of insecticide use for ECB prior to the introduction of Bt corn. Attributing any observed changes in insecticide since 1995 to the introduction of Bt corn is necessarily problematic for several reasons. First, insecticides are typically used for control of several target pests. USDA pesticide use surveys do not report insecticide use by target pest, which means that isolating insecticide use targeted solely to the ECB is impossible. Second, insect populations by their nature are highly variable from year to year, which makes trends difficult to discern. Finally, the introduction of new products and development of resistance in insect populations to older products also drive shifts in insecticide use. Five insecticides are currently recommended for control of ECB: Bt (foliar spray), chlorpyrifos, permethrin, lambdacyhalothrin, and methyl parathion. With the exception of foliarly applied Bt products, these insecticides are typically used for several target pests, including cutworms, rootworms, armyworms as well as the ECB.

Comparing 1995, the year before Bt corn varieties were introduced, to 1999, the use of these five insecticides declined. Of interest is the decline in use of four of the insecticides between 1995 and 1999: chlorpyrifos (-2%), permethrin (-1%), Bt (-1%), and methyl parathion (-2%). The decrease in the use of these insecticides may be due to the introduction of Bt corn varieties and the resulting reduced need for sprays targeted at the ECB. However, several other explanations are also possible to explain the change from 1995 to 1999. Lambdacyhalothrin was introduced in 1996, primarily for the treatment of cutworm in-season, which has likely displaced some of the use of the other insecticides, such as chlorpyrifos. Adult corn rootworm beetle populations have reportedly developed resistance to methyl parathion in some areas, which may account for part of the decline in its use. Finally, these reductions may be due to a decline in insecticide treatments targeted at pests other than the ECB. It is likely that all of these factors have affected use patterns of these insecticides.

The aggregate reduction of the percent acreage treated with the five insecticides is 6%. Assuming that 3% of the change is attributed to the introduction of lambdacyhalothrin, implies that a 3% decline occurred as a result of changes in the target pest complex, including ECB and other pests. For analytical purposes, it is assumed that one-half of the decline in the usage of the four insecticides (1.5%) was due to the introduction of Bt corn, implying over 1 million fewer acres sprayed for ECB.

TECHNOLOGICAL CHALLENGES IN AGRICULTURAL BIOTECHNOLOGY

Technological change in agriculture has been the foundation of economic growth and development. By raising yields and continually increasing the number of people who can be fed by the output of one worker in agriculture, change in agricultural technology has permitted an increasing proportion of labour and other productive resources to be devoted to non-agricultural production. Plant and animal breeding and chemical and mechanical technology plus improved husbandry have caused continual structural change in farming. These technological stimuli have arisen as a series of overlapping waves to create a process of technological change, the momentum of which looks likely to be maintained by intensified adoption of information and biotechnologies.

As a result of the cumulative processes to date, farms have become larger and more specialized, and farmers are more highly trained and have substituted machines for animal and human power; farming in the 'West' has become one of the most capital-intensive of industries. At the same time agriculture has become more dependent upon industry, and even more industrialized-the intensive production of eggs, broiler chickens, horticultural products and pigs has many characteristics which belong to industrial production rather than to traditional farming. Goodman *et al.* talk of 'the industrial appropriation of the rural processes'.

Processes of marketing farm products have passed off the farm to be performed by sophisticated distribution and retailing sectors which increasingly dictate details of production to the farm sector, while at the other end of the chain an increasing proportion of inputs is supplied by the chemical, pharmaceutical and engineering industries. The process of 'industrializing' agriculture has been transferred to the less-developed countries (LDCs) and is epitomized by the so-called 'Green Revolution'. This has entailed expansion of irrigation systems using modern pumps, engineered dams and canals, plus the increased use of inorganic fertilizer, insecticides, herbicides, fungicides and power machinery; although at its heart has been the 'old' biotechnology of breeding new cereal varieties which give high yields when supplied with chemical inputs. Because much of the impetus for this 'revolution' has been western technology and farming know-how, particularly that of the USA, there are those who are critical of what they see as the LDCs' increasing technological dependence upon the West, its industries and research institutions. What is undeniably the case is that technological advance has been faster, and public and private expenditure on research in western agriculture greater than in LDCs taken as a whole.

Thus exportable surpluses from the West have increased with generous public support, while many LDCs have had to resort increasingly to food imports to meet their needs. While at this stage it is difficult to foresee exactly when, and on what commercial scale, the new agricultural biotechnology 'revolution' which is now brewing will have its effect, most commentators agree that it will intensify the industrialization of agriculture, and that it will increase the technological dependence of most LDCs upon western firms.

It may cause considerable disruption to the economies of some countries and will cause farming operations to diverge increasingly from the pattern associated with the country yeoman. Rural life in many more remote farming areas will be further threatened by the continued deterioration in the economics of ruminant livestock farming on upland and low-productivity pastures. It is these redistributive and structural effects that are the central focus here rather than any attempt to estimate the scale and speed of uptake of specific biotechnology inputs and processes. Although prospective biotechnological change in agriculture is not different in many fundamental respects from other processes of technological change, in that it will operate through the provision of new or modified inputs (seeds, hormones, etc.) and the creation of new markets in industry, it does raise new issues.

Are there dangers from releasing life forms which are engineered using new methods which are quite different from those associated with varieties developed by traditional plant-and animal-breeding methods? Even if the dangers can be objectively assessed and turn out to be minimal, will what may be perceived as man-made and therefore 'unnatural' products be acceptable to society as a whole?

AGRICULTURAL TECHNOLOGIES AS EXPERIMENTS

Farmers have continuously, and quite consciously, experimented in their fields with crops. Most observers have overlooked this process, emphasising instead innovations that have originated in laboratories and field stations of research organisations and corporations established specifically for that purpose. Yet, 'probably, the total genetic change achieved by farmers over the millenia was far greater than that achieved by the last hundred or two years of more systematic science-based effort'.

In suggesting that developing country farmers were keen to utilise new technology if only it were available to them, a view deemed radical at a time when developing country farmers were visualised as both backward and conservative, The odoursSchultz implied that the technology they were employing was not changing. Similarly, Jackson and Ford-Lloyd in suggesting that artificial selection by humans was 'probably of largely unconscious nature', seem to devalue the contribution of early farmers in developing crop species.

Probably, Cox and Atkins are right to suggest some traits were selected for consciously, others by default. Nevertheless, the fact that seed selection has been practised all over the globe for millennia supports the view that one does not need a rigorous understanding of genetics to recognise value in such practices, and as Biggs and Clay point out, farmers work in continually changing environments. Without experimenting with crop varieties, most of the plants which today we recognise as crops would never have evolved to their current form.

DIVERSITY IN DECLINE

Since the mid-nineteenth century, an increasing amount of research into plant breeding has taken place in organisations set up partly for that purpose. As plant breeding became more successful, producing cultivars matching the needs of an increasingly market oriented agriculture (expansion of which was assisted by breeding), so there began a decline in the genetic diversity of crops in the field. Holden *et al.* suggest that crop genetic diversity probably peaked around the middle of the nineteenth century.

The double bind which faces those who have sought to replace landraces, or diverse crop mixtures, with uniform 'improved varieties' is that the genetic diversity on which their activity depends may be lost as a consequence of their very success. This is one way in which 'genetic erosion' occurs.

The value of landraces for the future of agriculture was recognised by two German scientists, von Proskowetz and Schindler, at the end of the nineteenth century. Nikolai Vavilov then laid the basis for work on centres of origin and diversity of crop plants, suggesting that the genetic diversity of a given crop coincided with its centre of origin.

An exact correspondence has been refuted, and today Vavilov's localised centres are discussed in terms of centres and non-centres which cover significant geographical areas.

However, the idea that some areas of the world exhibit great diversity for particular crops continues to be widely accepted, and is of crucial importance to breeders today as genetic erosion gathers pace, even in these areas.

Loss of diversity has taken place both between and within species, and as a result, it can be argued, in the range of techniques employed in agriculture: 'the mosaic appears to be disappearing'. As Wilkes notes, a shift from growing crops for consumption to growing crops for sale has resulted in:

the dropping of many legume crops (20–35 per cent protein) which have been classic protein sources in the human diet for more fertilizer/water-yield responsive cereals (7–14 per cent protein). The current situation is a rapid reversal of eight millennia of *in situ* evolution where the cultivators ate the crops they cultivated.

Today, of more than 200 000 plant species currently known to scientists, only 30 are used to supply humans with the bulk of their food. If one was to compile a list of the top five crops in every country, only 130 species would be named. It should come as no surprise that there are many 'unknown' crops ideally suited for production and consumption purposes in the regions where they have been selected.

Of those crops that are widely grown, the extent of diversity exhibited within the species is also narrowing. This entails risks in terms of the food supply's vulnerability to catastrophic losses, though as with many other ecological problems, quantitative estimates of such risks are subject to great uncertainty.

Even if the probability of such epidemics occurring were low (which we do not know), they may have severe consequences: 'Loss of genetic diversity – silent, rapid, inexorable – is leading us to a rendezvous with extinction – to the doorstep of hunger on a scale we refuse to imagine'. Given the prevailing uncertainty, it is equally valid to depict optimists as complacent as it is to brand the authors of the above words as scaremongers. Such is the state of knowledge about the system which feeds many of us.

Estimates as to the extent of genetic erosion tend not to take on a reliable quantifiable form, partly because by the time the problem began to be addressed, much diversity had already disappeared. Although the loss of varieties is nothing new, it is the pace of recent changes that is without precedent.

Since World War II, for example, virtually all of the local wheat varieties in Greece, Italy and Cyprus have been abandoned and most of the indigenous sorghum (*Sorghum bicolour*) races of South Africa disappeared after the release of high-yielding Texas hybrids.

But if the fact that genetic erosion is occurring is beyond doubt, the manner in which it occurs is disputed. Even those who emphasise the contributions made by traditional agriculture through crop selection have tended to fall into the trap of believing that the use of modern varieties always implies loss of traditional varieties. That they do so is surprising given that many seek to highlight the wisdom of indigenous practices. Why would such wise cultivators simply discard varieties they have so carefully selected and take a leap into the unknown? The work of Brush and others suggests that in some cases they do not.

The dismissal of *in situ* conservation is based on an extrapolation from temperate or from atypical regions of the tropics. This reasoning rests on conventional ideas about modernisation, which envision a linear replacement of earlier stages of agricultural development by more advanced ones... we should re-examine our assumptions about the loss of diversity, thus opening the plausibility of on-site conservation.

How new varieties are introduced and used in agro-ecosystems varies from place to place and from farmer to farmer. It is not impossible for modern varieties to augment diversity, or lead to the production of novel populations. Where this occurs, it reflects *diversity's value in use*. Even where replacement does occur, the value of discarded varieties may subsequently become apparent.

The Risks of Genetic Uniformity

Why is genetic erosion a problem? In traditional agricultural systems, epidemics are rare, unnoticed, undocumented, or all three. This is partly due to the genetic diversity, and hence, flexibility retained in the agricultural system. Genetically uniform systems lack this flexibility.

The more uniform the crop over space and time, and the less frequent are environmental conditions deleterious to particular pests and pathogens (they are most infrequent in the tropics), the more likely it becomes that an epidemic will occur, and occur over a large area. Modern agricultural systems are characterised by their uniformity. Ecologically, they are artificial in the extreme, and the level of inter- and intra-specific diversity is, in terms of the system's ecological vulnerability, worryingly low.

The possibility that pest epidemics will emerge increases due to the strong selection pressure exerted in favour of pests which can overcome any resistance that the crop may have had. In 1970, nearly a quarter of the US maize crop was wiped out by southern corn blight owing to the uniformity of the cytoplasmic DNA of hybrids making each equally susceptible to the fungus *Helminthosporium maydis*.

A subsequent National Academy of Sciences investigation concluded that, not just in maize, United States agriculture was, 'impressively uniform genetically and impressively vulnerable'.

Among those alarmed by developments was Jack Harlan, whose father Harry had warned as long ago as 1936 that: When new barleys replace those grown by the farmers of Ethiopia or Tibet, the world will have lost something irreplaceable. When that day comes then our collections, constituting as they do but a small fraction of the world's barleys, will assume an importance now hard to visualise.

The significance of Harry Harlan's comments lay in his awareness that genetic variation is what breeders require to develop new varieties, and lost varieties may contain alleles of genes whose importance cannot be estimated in advance. Well before Harlan made these observations, the same uniformity had led to:

- The potato famine in Ireland;
- The French wine industry being struck by powdery mildew in 1848;
- The destruction of the coffee industry in Ceylon (then the world's largest producer) in the late nineteenth century; and
- Two wheatless days a week in the US in 1917 due to red rust.

Since the corn blight episode, uniformity has led, amongst others, to problems with rust in the UK wheat crop in 1972 owing to widespread planting of two wheat varieties (supposedly resistant to rust); orange wheat blossom midge in the UK in 1993; rice blast in South Korea in 1980 when 75 per cent of rice lands were sown with modern varieties of similar pedigree (forcing the country to import rice; Russian wheat aphid in a number of countries including the US and growing problems in 1990s US and the UK with the fungus responsible for the Irish potato famine, *Phytophthora infestans*.Of all these episodes, only the case of potatoes in Ireland can be said to have created a famine. Yet the potential exists for widespread crop losses, and in a world where global food stocks are shrinking, the likelihood of such events occurring in the future may be increasing.

The standard approach to dealing with vulnerability associated with genetic uniformity has been to breed for pathotype specific (vertical) resistance to prevent losses due to a particular ecotype of a pathogen. For the last fifty years or so, this 'disadaptive breeding' has been the approach of choice for formal agricultural research. Yet vertical resistance exerts the strongest selection pressure in favour of 'matching' pathotypes, limiting its useful life.

Because the search for vertical resistance makes the degree of horizontal resistance difficult to detect, the level of horizontal resistance tends to be lower than is desirable, resulting in a boom-bust cycle.

This is the so-called vertifolia effect. Breeders partake in what has become known as a varietal relay-race in their efforts to pre-empt the devastation of crops by pests and pathogens.

In situations where vertical resistance is matched, or where vertical resistance cannot be found (because it may not exist), the use of pesticides is unavoidable.

The use of pesticides, in turn, has led to development of new ecotypes of old pests and pathogens which are resistant to chemical attack, as well as the emergence of new pests and pathogens that thrive in the absence of their natural.

When spraying to kill some pests, particularly those whose vectors move easily between fields, farmers create positive benefits for themselves, and a positive externality for other farmers in that they help to contain populations. Yet they may also incur costs to themselves in terms of health impacts, and lost productivity as a result, as well as two types of negative externality. The first relates to environmental pollution.

The second relates to the fact that pest resistance builds up at a rate determined in part by the volume of the active ingredient used. All farmers are adversely affected in the medium term through yield loss owing to existing resistance being overcome.

Uniformity is, therefore, a problem with at least two related dimensions; vulnerability, leading at best to market instability and loss of earnings, and at worst, to widespread hunger; and associated environmental pollution. Alternative strategies to vertical resistance include:

- Breeding for pathotype non-specific (horizontal) resistance – horizontal resistance, though more difficult for plant breeders to achieve, is rather more durable. Those favouring vertical and horizontal resistance can be said to be adopting, respectively, Mendelian or biometrical approaches to the problem. The problems associated with vertical resistance are increasingly well recognised and a shift is occurring towards horizontal resistance strategies. However, horizontal resistance strategies are not without problems of their own, one of which is that it is not known how effective horizontal resistance strategies will be in the absence of agrochemical use.

- Diversifying the varietal make up of a crop in any given region – by planting varieties incorporating different vertical resistance genes in neighbouring fields, the possibilities of disease epidemics occurring are reduced. The method is most effective where an early variety is the source of reproduction for a pathogen of a late variety. Similarly, where a parasite is known to reside in a particular region in a given season and then spread in subsequent seasons, growing varieties with different resistance genes along its path may slow down, or prevent, development of related disease. In a slightly similar vein, Wilkes talks of the use of 'genetic firebreaks', bands of diverse germplasm employed to limit the progress of any epidemic.
- Varietal rotation. Varieties incorporating different resistance mechanisms are grown once every few years and rotated with others.
- Incorporating, in a given field, diversity within varieties, between varieties, or between species. At one end of this spectrum are multi lines and mixtures, in which different vertical resistance genes are either backcrossed into a single genetic background or used in distinct varieties. At the other end are the diverse systems in home gardens, and of permaculture and agroecology, where both inter- and intra-specific diversity are encouraged.

One can appreciate that these different recommendations require different responses from breeders, farmers and policy-makers. The methods vary in terms of their genetic flexibility.

This may be a consideration that increases in importance if climates change quickly (in historical terms), causing 'environment erosion' of any resistance (vertical or horizontal) present in monocultures. Furthermore, use of diversity within the agroecosystem may offer the best prospects for eliminating the use of synthetic pesticides, though some advocates of horizontal resistance do so on the basis of its performance in this respect.

Risks associated with a narrow genetic base are reduced by widespread agrochemical use (with attendant consequences for the environment), and by substantial investments by public and, increasingly, private organisations whose aim it is to keep hold of the batton in the varietal relay-race, either through developing new chemicals, or new varieties.

This places a considerable burden on these organisations to continually come up with the goods in the face of unexpected and unannounced attacks from pests and pathogens, as well as (possibly) changing climate. As pathogens overcome vertical resistance conferred by individual genes or become immune to chemical attack, and as genetic erosion continues in crop plants, weedy relatives of crops are assuming greater importance. Harlan has been moved to comment that 'wild relatives stand between man and starvation.' The suggestion that genetic erosion is a problem does not inevitably lead to the recommendation of a re-direction of research in favour of practices that maintain, enhance, or encourage diversity *in situ*. Many advocate a more comprehensive, better described, and better co-ordinated system of *ex situ* conservation in gene-banks. Those who support this approach have confidence in *ex situ* conservation to provide adequate back-up for a breeding system which they implicitly trust to overcome any future problems.

This position is increasingly being undermined by well-documented limitations of gene-banks, leading to growing recognition that *ex situ* storage of germ-plasm is inadequate of itself to ensure a secure genetic foundation of agriculture into the future.

This has led some to advocate acknowledgement, and some form of reward system, for the role played by various farmers and communities around the world in maintaining crop genetic diversity *in situ*. Within this school of thought, there is a difference in emphasis between those who advocate the conservation of genetic diversity *in situ* for use in future breeding programmes, and those who recognise diversity as a desirable property of agricultural systems in itself.

The former view lends itself open to criticism on grounds that it can be taken to imply that *in situ* diversity would inevitably disappear unless prevented from doing so. But the latter view, exemplified by the work of Brush and others, immediately suggests a need for reappraisal of the objectives of formal agricultural research. If diversity is of value in use, why do researchers not base their investigations more firmly on the incorporation of diversity at the inter- and intra-specific level? If farmers are capable of experimenting with novel germplasm, why do *ex situ* gene-banks fail to promote an active exchange of their resources with farmers?

External-Input Agriculture

It is often stated, indeed, it is frequently taken as axiomatic that modern high-external-input agriculture (HEIA), unashamedly uniform genetically, constitutes the only possible way to feed the worlds large and growing human population. There are many who believe that to retain genetic diversity, in whatever way, within the agricultural system necessarily imply an economic loss, or at least, a loss in production. Supporters of this view point to the achievements of the Green Revolution in Asia and parts of Latin America. Lipton and Longhurst write: 'history records no increase in food production that was remotely comparable in scale, speed, spread and duration'. This most probably is the case, yet the same could probably be said for the accompanying narrowing of the genetic base, and indeed, the 'scale', 'speed' and 'spread' factors each have negative ecological manifestations.

Many who discuss the Green Revolution tend to limit their number of counterfactuals to one, a continuation of the *status quo*. Is it to be understood that agricultural research could only work effectively if it concentrated on monocultures? Could production increases not have been achieved in any other way? Is it impossible to construct plausible counterfactuals paths for change which left agriculture less vulnerable in terms of its genetic inflexibility, less dependent on crop protection chemicals, and yet with the potential (in the sense that it meets the sufficiency criterion) to feed the world?

It seems almost heretical to write these words, but even on grounds of productivity alone, it may be true to say that (and this may require qualification with respect to ecological regions) agroecosystems incorporating greater species diversity are the best performers in terms of output per unit area and/ or economic performance. At the same time, they can offer a number of environmental benefits relative to HEIA (depending on the system) which may include energy conservation, reduced agrochemical use, soil conservation and resource use efficiency. Hence, there are three important, if for the most part obvious points to be made with respect to the position that HEIA is the only way to feed the world:

- Modern agriculture does not have, as one of its aims, feeding the world. It is principally geared to making profits;
- Farmers have a number of reasons for adopting one or other technology. Concern for global food supplies and their distribution does not weigh significantly in such considerations; and
- Nobody has, and nobody could, prove the truth of what is being stated. The choice with regard to approaches in agricultural research is far wider than is often understood, and suggesting that one or other is 'the best' (*let al*one 'the only') is likely to prove problematic, especially when so few of these approaches have been researched thoroughly.

There is reason to believe that more will be known about any given way of doing agriculture as more research is conducted upon it, raising the possibility for what we might term increasing returns to advocacy. In other words, if through gaining an early lead in the command of research resources, a particular approach improves its position relative to alternative approaches, it is more likely to command research funds in the future.

Whilst Plucknett criticises those who argue the need for research on marginal lands (as opposed to the most productive lands), Lipton, mindful of the over-simplifications implied by researchers in dividing land into advanced or backward areas, asks the question:

Are the 'lead areas', often no longer the most promising (especially after much research with a long period of diminishing returns), but rather the areas that have been favoured by AR [agricultural research] in the past, and continue to be favoured because that is what scientists know how to do best?

Two questions need to be answered; why has one particular way of doing agriculture come to be so widely embraced the world over; and what are, what has happened to, and what might be the alternatives?

Comparing Farming Systems

In order to understand why it is at least somewhat odd that a particular way of doing agriculture has come to be accepted so broadly, it is worth considering some of the factors that might impinge upon the decision to farm in whichever way. The problems encountered when trying to compare farming systems are many and varied. They relate to:

- *Location specificity* – the only satisfactory way to compare two farming systems is not available to us. We cannot compare different systems at the same time on the same piece of land. Because the quality of land will be affected by past activities, a comparison conducted sequentially on the same piece of land will introduce errors. If the comparison is not to take on the form of an experimental test (and it is generally appreciated that research station performance differs widely from that in the field) one faces the problem of finding working farms which are not only similar in agroecological characteristics.
- *Motivations for farming* – farming systems are part of wider social systems and are embedded within a social context. Not all farmers have the same motivations, particularly if farming is not the primary source of income. Even where income maximisation processes are of importance, they may still take on different forms. Many farmers will seek to optimise performance with respect to multiple criteria. Reintjes *et al.* consider productivity, security, continuity and identity to be important objectives, any combination of which could constitute the guiding rationale for farming. Problems that might arise from a desire to control (if this were possible) for variation in motivations include the possibility that choice of farming technique may be, indeed almost certainly is, endogenous to these motivations.
- *Managerial capabilities* – that one farm 'does better' than another in a given comparison could simply be regarded as proof of the superiority of one farmer's management skills. But it is difficult to dissociate the issue of management from that of the farmer's objectives, and indeed, those of the researcher. One might also ask whether the adoption of one or other farming system should itself be considered an issue in measuring management capabilities. Furthermore, the adoption of one or other system might *require* of farmers that their management capability improves. Perversely, this might raise questions as to the wisdom of farmers deciding to practice more managerially demanding styles of agriculture.

Different types of farming systems probably require completely different types of managerial skill, if that is the correct phrase. These considerations serve to caution us against treating technologies and techniques as isolated parts of what is actually a complex whole, in which the ability and resources available to manage the farm are often conditioned by the characteristics of the farm household. How the household functions as a unit, and how it interacts with the wider community, will affect the types of decision taken regarding farm management. *Duration of comparison* – farming has always been a risky enterprise, and although in different farming systems the principal source of those risks (and possibly where the burden of risks falls) is altered, it continues to be so. Issues of seasonality, year-to-year variation, and secular trends or even breakdowns in one or other variable in the system are of crucial importance.

Most of these can only be revealed through a comparison which takes the longer-term into account. On the other hand, such studies of a longer-term nature as have been done have taken the form of experiments which, due to the degree of control that is desired, tend to ossify the production systems under investigation. This prevents any evolution in the system that might be desirable or necessary over time. Given that, for many systems, design and its improvement are of supreme importance; such comparisons lose relevance over the longer term. *Differences in systems* – farming systems may be so different that they simply are not comparable in any meaningful sense of the word.

The sort of alternative systems considered above incorporate greater diversity on the farm, leading almost inevitably to a more diverse output than is characteristic of modern, highly-specialised farms.

How can one really compare the output of such different systems, particularly when one also considers the many other factors mentioned above which mitigate against any easy comparison? Research methodologies which use the manipulation of a restricted number of variables as the basis for comparing organic and high-input agriculture have probably missed the point of, for example, organic agriculture altogether.

They suggest an inability to appreciate that in such agriculture, what is sought is system integration through the interaction of components, and this is a goal rather than an instrumental means to a particular end. *The comparison to be made* – to the extent that a focus on one or more indicators inevitably fails to consider all the dimensions of the farming system, any comparison will be less than complete, and therefore prone to biased conclusions.

If such comparisons are attempted, it seems impossible to dissociate the comparison from the motivations of researchers. Additionally, if a comparison is made on the basis of chosen indicators, for the comparison to be a fair one, the farmers being compared should indicate an equal weighting of these indicators in their subjective assessment of farm performance.

Clearly, comparison of alternative farming systems is extremely difficult. Yet for all these difficulties, the homogenising tendencies of the expanding agro-food system are difficult to deny, still more so to resist. Why do so many farmers apparently chose independently to farm in a broadly similar way the world over?

Have the options open to them been inadequately represented? To state that alternative options have received inadequate attention from formal agricultural research is partly to answer the question, 'why are these options called alternatives?' Consideration of this issue is important in the light of the genetic vulnerability inherent in the prevailing practice of agriculture, and the environmental problems this creates.

Schools of Alternative Agriculture

As applied to agriculture, my use of the term 'alternative' is culturally biased. In some regions, modern high-input agriculture is simply unfeasible and 'alternative' agriculture is all that is available. Furthermore, several alternative movements have referred to the validity of indigenous technologies which have, it goes without saying, pre-dated such movements by some considerable time. Harwood comments that the roots of alternative agriculture are to be found in the discourse evolving at the turn of the twentieth century juxtaposing holism and reductionism.

Alternative agriculture's development parallelled that of modern agriculture, leaning on the insights of the growing number of soil scientists, and influenced by Darwin, who had a great interest in the biological attributes of soil. In the early twentieth century, several books were written which alluded to the complex interrelationships between components of the farming system that had been ignored or downplayed by agricultural scientists.

The work of King and Howard on highly productive systems of agriculture in Asia helped reinforce this message. However, alternative movements seem to have been severely affected by the Second World War. For practitioners of biodynamic agriculture, based on Steiner's anthroposophy, the impact came through ideological channels, as the National Socialists banned their activities.

Harwood cites a number of works which appeared in the 1940s and 1950s, a common theme of which was 'the increasing environmental harm and resource degradation brought about by 'modern' farming methods.

Yet, for Harwood, the 1960s were a transition period of 'narrow focus', which saw the rise of modern agriculture and the parallel demise of alternatives in agricultural scientific discourse.

Harwood's view is corroborated by the experience of different nations in the post World War Two era, during which the state has played a crucial role in the promotion of modern agricultural techniques.

Most interesting to note in this context is the way in which the prosaic critique by Rachel Carson was greeted by Harwood and his colleagues.

There was little or no debate during those years, in the biological sciences at least, on development direction. The success of current technologies was so overwhelming that it stifled serious debate of alternatives. The alternative farming 'schools' were practically non-existent and certainly in disrepute. I remember clearly graduate school discussions in 1964 about the 'crackpot Rachel Carson and her whistling in the wind' against the great benefits of DDT. In looking back on those heady years, I wonder about our arrogance and narrowness of vision.

ADVANCED AGRICULTURAL BIOTECHNOLOGY

Every year, millions of people worldwide die from starvation and nutritional deficiencies. Although these challenges remain significant, we are making some progress. Agricultural technologies have increased caloric intake on a per-capita basis in nearly every nation of the world. technological development — particularly in the area of agricultural biotechnology — is crucial if we want to further reduce malnutrition and starvation and meet the needs of growing populations.

Unfortunately, misperceptions about biotechnology have led to activist movements to slow down, if not eliminate, this critical technology via government regulation. In the United States, the advancement of biotech crops has slowed down because of attempts to regulate them differently than conventionally grown foods.

In Europe, antibiotech movements are much stronger, and international bureaucrats are pushing various international agreements and trade restrictions. While Western nations may be able to absorb the costs of such policies, people in the developing world will pay a higher price as many continue to suffer from starvation.

For centuries breeders have improved plants through selective breeding and crossbreeding different types of plants. Agricultural biotechnology — also referred to as "genetic engineering" or "genetic modification" (GM) — is a more sophisticated and precise means of modifying plant genes. Instead of transferring thousands of genes as traditional crossbreeding does, biotechnology enables the transfer of only a few select plant genes.

Breeders essentially cut the desirable traits out of one plant's DNA and paste them into another — a process called "gene splicing." In addition, undesirable traits can be targeted and removed. Using biotechnology, plant breeders can produce better crops.

BIOTECHNOLOGY'S CRITICAL BENEFITS

More affordable, higher quality food surely benefits people in developed nations, but those benefits pale in comparison to what biotech promises for the developing world.

Biotechnology promises to reduce world hunger and disease by improving local productivity by adapting crops to local climates and soils; increasing yield by making plants stronger and more pest-resistant; making plants more nutritious by creating plants with higher vitamin and protein content; and making produce more affordable on the world market. For example, Cornell University researchers are developing a rice variety that yields 20 percent to 40 percent more food per acre and a tomato that produces 48 percent higher yields per acre.

Biotechnology promises to play a critical role in reducing blindness caused by vitamin A deficiencies. Every year in Southeast Asia, one quarter of a million children go blind because of a lack of vitamin A in their diets. Worldwide, 124 million children suffer from vitamin A deficiency. Increasing the intake of vitamin A could prevent one to two million deaths per year and eliminate numerous health problems.

Poverty and lack of infrastructure make the traditional methods of vitamin consumption (a varied diet or supplements) impossible. Agricultural biotechnology, however, can help to address the problem of childhood blindness and death by ensuring that people get this nutrient in the key staple of their diet, rice. Researchers have spliced into the genome of rice, the beta-carotene gene from a daffodil to create "golden rice" — rice high in beta-carotene, which the human body converts into vitamin A.

This is among the most well-known examples of biotechnology's benefits. Countless other genetically engineered crop varieties are being developed to address similar problems of poor nutrition or productivity.

AGRICULTURAL BIOTECHNOLOGY IMPROVES PLANT SAFETY

Many biotechnology skeptics contend that such engineering is unsafe because it is "unnatural." Ironically, biotechnology is actually an improved method for plant breeding. Most of what we eat today has already undergone gene manipulation under the traditional breeding methods. The traditional methods and biotechnology differ in that traditional methods are cruder because they do not give researchers control over all the genes they crossbreed.

Some desirable characteristics might result with such methods, but undesirable ones might as well. It could take a long time of continual breeding to eliminate the most problematic traits and focus on the best ones.

Compared to these more "traditional" methods of hybridization, genetic engineering is a wonder of precision and accuracy. Genetic engineering enables scientists to pinpoint the genes that express a desirable characteristic in one plant and to insert those genes into the genome of another plant.

Scientists mark the insertion point so that they know where these novel genes are located in the genome of the new plant. While there are several different methods of genetically engineering plants,

CHALLENGES TO CROP GENETIC DIVERSITY

Farmers have traditionally depended upon their own skills and resources to develop the crops that they need. The process began with the earliest domestication of wild species and has continued through the careful selection of plant materials adapted to varied (and often changing) growing conditions and preferences. The result has been a complex and continually evolving collection of local crop varieties, often referred to as landraces, that reflects interactions with wild species, adaptations to changing farming conditions, and responses to the economic and cultural factors that shape farmers priorities.

The richness and range of the diversity of these landraces is now under threat because of the changing nature of agricultural production. One important factor is the widespread adoption of modern varieties (MVs) that are the products of formal plant breeding. These MVs often provide yield increases and other advantages that result in their being sown over large areas. Technological change, such as the use of fertilizer and irrigation, also lowers the demand for landraces adapted to marginal growing conditions.

In addition, farm production is increasingly market- oriented, making farmers less inclined to select for crop characteristics that once were important for local customs and culture. If the adoption and use of MVs and other agricultural technology brings significant benefits to farmers, why should we be concerned with the loss of landraces and the preservation of crop genetic diversity? There are several important reasons.

First, many advances of modern plant breeding have been possible because of the wide range of genetic source material provided by landraces. The very success of modern plant breeding now threatens the source of genetic diversity on which further progress depends, as farmers find it less rewarding to maintain the diverse mixture of landraces developed by their ancestors.

Second, the widespread use of MVs raises questions about the stability of crop production and the threat of disease or pest attack. The threat to stability does not necessarily derive from the MVs themselves. MVs are more uniform than landraces, which may increase their susceptibility to pests or pathogens, but many MVs owe their acceptance to the fact that they are more resistant than the varieties they replace.

The principal threat to yield stability from MV use is the increasing uniformity, and continuous cropping, that their use engenders. Large areas planted to a single variety are potentially a cause of concern, no matter what the source of the variety.

In addition, the preservation and utilisation of crop genetic diversity is of particular importance to the more marginal, diverse agricultural environments where modern plant breeding has had much less success. Farmers in these areas tend to be poorly served by public research and extension systems.

These areas are often centres of diversity for many crop species, but increasing poverty is forcing many of these farmers to place more dependence on non-farm sources of income, with consequent reduction in their capacity to grow and maintain the range of local varieties they have been accustomed to manage.

PLANT GENETIC DIVERSITYIN THE PERUVIAN ANDES

Farmers in the Andes of southern Peru depend on both potatoes and maize as basic staples. A study in the area of Paucartambo illustrates the complexity of genetic diversity maintenance for these crops and provides examples of the criteria that farmers use to select their varieties. Potatoes are the most important crop in the region. Local potato varieties are classified first by use category, divided among those appropriate for boiling, soup-making or freeze drying.

Further sub-classification is done on the basis of morphological differences, and nomenclature reflects a rich cultural symbolism. Most farmers manage between 3 and 6 separate potato fields in which a mixture of varieties of a single use category are planted together. Women are in charge of selecting planting material; for any one field, they specifically select only a few of the most important varieties and then make a representative selection of the remaining material to provide the desired diversity.

Observations in fields of boiling potatoes revealed a mean of 21.1 varieties, with a standard deviation of only 5.1, indicating a remarkable consistency in maintaining field diversity.

Maize is also classified by the farmers of Paucartambo into use categories, distinguishing among boiling and parching. In contrast to potatoes, sub-classification of maize varieties is based more on specific choices for use, yield, and time of planting. Also in contrast to potatoes, maize varieties of different use types may be mixed in the same field. The average number of varieties per field is much lower than for potatoes, and the women who select the seed for planting make much more precise choices.

Finally, the maintenance of a wide and evolving range of local crop landraces is threatened by the advent of intellectual property protection for crop varieties, accelerated by the formation of the World Trade Organization. The increasing application of plant breeders rights has several implications for plant genetic diversity. For a new variety to be legally protected, it must be subject to very precise description, including the requirement that it be distinct, uniform and stable (DUS). This is a disincentive to the promotion of intrinsically diverse landraces or of varietal mixtures.

An additional debate concerns farmers privilege, the ability of farmers to save seed of a variety, to exchange it with neighbours, and to adapt it to their own growing conditions. These practices could be challenged by seed companies when crop varieties are sold under strict legal protection. It is even possible to envisage situations where varieties that originated in farmers fields may be legally protected and then denied to the farmers responsible for having developed them. Hence the advent of plant variety protection lends added urgency to the search for solutions to the conservation of plant genetic diversity.

INNOVATIONS AND INITIATIVES

There are several important strategies that address the conservation of plant genetic diversity.

Genebanks

The conventional solution to the conservation of plant germplasm has been through genebanks. These are based on collections of genetic material, from centres of crop origin and elsewhere, that are stored in controlled conditions and periodically regenerated. Genebanks have been important sources of material for plant breeding programmes and other research activities, and will continue to be a basic element in conservation programmes.

Genebanks are not without their problems, however. Although some of the larger genebanks contain impressive numbers of accessions, the costs of characterising and cataloguing genebank material are considerable. Only a small proportion of genebank material is actually used by plant breeders, in part because of inadequate characterisation of accessions. In addition, farmers usually do not have easy access to the materials they have donated. The control of genebanks is also an issue of debate; there are concerns that a country that hosts an important international collection can deny another country access for political reasons.

These issues of control are beginning to hamper efforts at germplasm collection, as countries are increasingly concerned with the possibility that their local landraces may be collected, utilised commercially, and even legally protected, without any compensation. More resources are needed to support national plant genetic conservation efforts, and more attention needs to be devoted to making these collections relevant to farmers interests and capacities. In some cases, national genebanks are being redirected to encourage more farmer participation in their development and utilization.

Crucible Group

The Crucible Group is a collection of 28 individuals including representatives of grassroots organisations, agricultural researchers, intellectual property specialists, trade negotiators and agricultural policy analysts from South and North which met to discuss the issue of the intellectual property protection of plant genetic resources.

Not surprisingly, the group s discussions produced varied points of view, but also developed a consensus on a number of important issues. Some of the group s conclusions include: The Crucible Group stresses the primacy of specific national conservation strategies for plant genetic resources that invite the participation of local communities as well as private companies. Holders of ex situ germplasm collections should develop equitable partnerships with indigenous and rural societies and make their collections available to them.

Under the pressure of possible exclusion from an encompassing global trade agreement, many countries feel pressed to adopt some form of intellectual property protection for plant varieties.

The Crucible Group concurs that compulsion is inappropriate and that countries, have every right to protect their environment and the well-being of their peoples if they feel that trade rules threaten their security.

In Situ Cconservation

An option that is attracting increasing attention is the possibility of carrying out genetic resource maintenance in the field. This is often referred to as in situ conservation, in contrast to the ex situ alternative of genebanks. Arguments in support of in situ conservation include the importance of recognising the roles of both environmental factors and farmer intervention in landrace development.

In areas of high crop genetic diversity, landraces often evolve through crossing with wild or weedy relatives, and farmers play a crucial role in selecting and adapting new material. It is this process of landrace development that in situ conservation attempts to enhance and preserve. It is also argued that in situ conservation programmes would provide more opportunities for community conservation, where farmers would be better able to recognise and utilise the plant genetic resources that are available in the community.

But there are several decisions that must be made in designing in situ conservation projects. First, because landraces are not static entities, decisions have to be taken with regard to the nature of human intervention in the selection process. Whose criteria are to be used in the selection and adaptation of new materials? Are local farmers criteria the only ones to be applied in deciding what is conserved, or should scientists interests also play a role in determining the direction of conservation? In addition, farmers generally seek germplasm from other sources to complement their own landraces. To what extent should this be allowed or encouraged in an in situ conservation project?

To what extent is the objective to build a fence around an area of genetic diversity in order to protect traditional crop development processes from outside influences, and to what extent is the objective closer to that of community development? In the latter instance, resources and information are provided to farming communities to empower them to make more informed decisions about the management of local varieties and the utilisation of MVs.

In situ conservation projects will undoubtedly approach these decisions in a variety of ways. At this stage it is important to encourage a range of in situ programmes and to learn from experience. The International Plant Genetic Resources Institute (IPGRI) in Rome is currently developing a large multi-country collaborative project to test and develop techniques for in situ germplasm conservation.

Farmer-Participatory Plant Breeding

There are also a number of efforts under way to encourage a wider scope for farmer participation in formal plant breeding. Possibilities include greater farmer representation in priority setting for crop breeding programmes, more explicit attention to the crops and varietal characteristics of importance to these farmers, the transfer of significant aspects of plant breeding research to farmers fields, and the enlistment and training of farmers to take a more active part in the variety testing and selection process.

There is a growing literature on methods to encourage farmer participation. The innovations include rapid rural appraisal techniques to understand farmer varietal preferences, the organisation of various types of adaptive on-farm research to test varieties under farmers conditions, the wider use of landraces in formal breeding programmes, and the establishment of mechanisms for contact between farmers and experiment station personnel. Some plant breeders see the possibility of an integrated system that incorporates the strengths of both formal and informal plant breeding techniques.

Farmer Participation in Gene Banks

The Plant Genetic Resources Centre/Ethiopia (PGRC/E) is involved in several activities that involve farmer participation in conservation activities. The Unitarian Service Committee of Canada helps support a PGRC/E project in which materials collected from nearby areas are given to farmers to plant and to carry out simple mass selection to improve their characteristics. Farmers select plant types on the basis of characteristics important to them, including pest or disease resistance or maturity. Farmers receive assistance from plant breeders in this activity, and PGRC/E scientists establish standard descriptor lists for the materials.

Farmers evaluate their selections in experimental plots and compare performance and yields with the original seed stock. The plots are also used for on-site maintenance of landraces. In another project, the PGRC/E is collaborating with Debre Zeit Research Centre to maintain indigenous material of tetraploid wheat. Wheat germplasm collected by PGRC/E is tested and multiplied by breeders and then provided to farmers for further multiplication and selection. This allows farmers to experiment with landrace lines while the indigenous populations are maintained in the genebank.

KNOWLEDGE GAPS

Although there is undeniable evidence of the erosion of crop genetic diversity, and several innovative responses have been developed, there are important gaps in knowledge that limit our capacity to decide among the various alternatives. Some of these gaps involve our technical understanding of the nature of biodiversity, while others are concerned with our understanding of its socio-economic correlates. Appropriate measures for diversity still need to be developed in order to better characterise the current situation and to evaluate changes in the future.

Farmers crop variety classification systems are one place to start, but we do not know how these correspond to actual genetic differences. With few exceptions, there is little research that correlates the variation in folk taxonomy with genetic differences.

The capacity to evaluate genetic material in the laboratory is growing rapidly, but these are still expensive techniques, and more robust markers and measures are required to follow the progress of genetic resource conservation.

When better measures of genetic diversity are devised they will contribute to a clearer understanding of what exactly needs to be conserved. Currently, there are only very general ideas about what portion of a plant population needs to be maintained in order to conserve particular genetic traits.

This information is crucial to the efficient design of in situ conservation projects, for instance, in order to know if efforts might be limited to a fewfarmers fields, or should instead sample more widely from an agroecosystem.

More study is also required to understand the causes of plant genetic erosion. In some cases, the introduction of MVs has led to the replacement of large numbers of local varieties. On the other hand, several studies of varietal adoption in centres of crop diversity have shown that MVs often complement, rather than replace, local varieties.

The availability of inputs and crop management techniques that reduce the need to tailor varieties to difficult growing conditions has also contributed to the demise of many local varieties.

The cultural value placed on crop diversity and local selection techniques is also diminishing in many areas, and the skills that have contributed to landrace evolution are consequently disappearing. As farmer variety selection skill is a threatened resource, there is also a need to understand how to conserve and enhance it.

Although farmers deserve credit for landrace development, there is little knowledge of what the genetic consequences of their selection techniques are, or what specific effects these have on enhancing diversity. Farmers select materials for practical reasons that may not always be compatible with the maintenance of genetic diversity.

Very little interdisciplinary study has been devoted to understanding the biological outcome of the application of indigenous technical knowledge and skills in variety selection.

There is also little understanding of whether attempts to improve local selection capacity should focus on individual farmers or on communities. Indeed we know very little about how varieties, and information about them, are exchanged within communities.

POLICY IMPLICATIONS

The challenges of crop genetic resource conservation highlight once again the dilemma of balancing between development and conservation. The development contribution of modern plant breeding weakens diversity by replacing many local varieties, and thus threatens the possibilities of further progress.

Areas of high genetic diversity, on the other hand, are often distinguished by their poverty and the declining viability of their agriculture, and are desperately in need of innovation and change.

The dilemma is evident in choices of conservation strategies as well as in the design of development programmes. The identification of an optimum mix of development and conservation initiatives is one of the most difficult tasks to face policy makers in the next decade, and the necessity to develop location-specific strategies adds to the complexity of the challenge.

Much more effort is required to develop adequate analytical tools to enable policy makers to explicitly address the trade-offs and consequences of particular decisions.

The Integration of Farmer and Formal Plant Breeding

It should be possible to strengthen farmer breeding efforts by linking them to elements of the formal system. If the tools and methods of formal plant breeding were made available to traditional systems, an integrated plant breeding model could be the result.

The choice of appropriate plant genetic resource conservation strategies also requires clear policy decisions about the appropriate mix of public, commercial and voluntary contributions to variety development and seed distribution.

Commercial seed enterprises are now replacing many public or parastatal seed operations, and are also making an increasing contribution to plant breeding and variety development. In many instances, seed companies will be able to respond more effectively to farmers needs than the public sector.

But the commercial sector will not be likely to address many of the minor crops or special growing conditions that are important to resource-poor farmers, nor will they be likely to play a prominent role in conservation activities.

Adequate legal protection for commercial varieties should be balanced against the assurance of farmers privilege to save and adapt their own seed.

Too much current seed policy advice focuses on the negative experience of public sector seed production and ignores the new set of responsibilities that public seed policy and public research will have to assume in order to assure more equitable agricultural development. As is clear from the discussion on in situ conservation and new approaches to plant breeding, plant genetic resource conservation will increasingly involve farmer participation. This will require innovative interaction among public sector and voluntary or community organisations. Policy makers need to devote more attention to fostering long-term, productive collaboration between public agricultural research and organisations active at the community level.

A particularly challenging feature of conservation programmes is their necessarily long-term nature. Both national policy makers and external donors who wish to support conservation programmes must assure that funding is available for an extended period to include the necessary research, training, and implementation.

Piecemeal efforts at genetic resource conservation that do not feature a long-term commitment will accomplish little. This puts a premium on careful planning and coordination among participating agencies, and implies a significant re-orientation to the way that donor funding for agricultural development is currently managed.

Finally, good policy always depends on good information, and this is particularly true for genetic resource conservation. Despite rapid progress, serious gaps in our knowledge are likely to constrain the informed management of plant genetic diversity for at least the next decade. The issues demand interdisciplinary collaboration among social and biological scientists.

They also require increasing collaboration between researchers and the members of grassroots development initiatives. Policy must direct national research and academic institutions to give priority to collaborative research on conservation issues, and should take responsibility for establishing appropriate fora where different perspectives can be presented, debated, and synthesised.

BIOTECHNOLOGICAL DEVELOPMENT IN AGRICULTURE

In a recent review of agricultural technology, the Office of Technology Assessment (OTA) of the US Congress defined biotechnology to include 'any technique that uses living organisms or processes to make or modify products, to improve plants or animals or to develop micro-organisms for specific uses-it focuses upon recombinant DNA and cell fusion technologies'.

Longworth concurs with this definition, with the significant addition of tissue-culture techniques, an aspect of biotechnology which has great economic, and therefore social and political, potential impact. Despite debates about whether these or any other definitions are adequate, they do touch on the main aspects of biotechnology which deserve to be considered.

The aspect of this new biotechnology which most captures the imagination and stirs the greatest controversy is gene splicing or recombinant DNA techniques, which inspire researchers to consider the possibilities of producing reproducible animals and plants markedly different from current existing species, and referred to as transgenic species. Already in higher animals and plants recombinant DNA has produced transgenic forms which are being commercially exploited. Their agricultural significance is, however, so far limited, and commercial application is concentrated in highly profitable pharmaceutical and horticultural markets.

Genes have been introduced into several animal species which alter their protein synthesis to enable transgenic sheep to produce insulin in their milk, and rabbits to produce interferon.

A completely different application has originated in Denmark for salmon, where it has proved possible to introduce germplasm which enables the salmon's physiology to handle heavy metals, which are normally toxic, so opening up new locations for farm fisheries.

Much current research is directed to conferring disease immunity on animals, and holds out the prospect of widespread commercial application. In plants one achievement has been the transfer of genetic resistance to antibiotics in the petunia, and another has been the introduction of storage-protein genes from French bean plants into tobacco plants.

As yet commercial progress with recombinant-DNA technology in plants appears limited, but extensive opportunities beckon, particularly because plant research is less restricted by the ethical and animal-welfare concerns which apply to research on transgenic animals.

The most important commercial developments based on gene splicing have so far occurred with genetically much simpler microorganisms, and it is with these that the greatest short-to mediumrun commercial potential lies.

Already genetically engineered micro-organisms are producing a variety of hormones, vaccines, enzymes and other proteins. Important examples are the production of insulin, vaccines for neo-natal diarrhoea in calves and piglets, and the bovine growth hormone BST identical to that produced in cows which can stimulate a 20 per cent increase in milk yield.

BST is already a source of problems for legislators in the European Community, and has provoked strong reactions from the media and milk consumers. As far as larger animals, and cattle in particular, are concerned, it is developments in embryo transfer and in many processes for manipulating reproduction which hold out the prospect of continuing increases in yields, feed conversion efficiency and general economic efficiency. Already, apparently over 1 per cent of dairy calves in the USA are from embryo transplants, despite the high costs still associated with this procedure.

Widespread adoption of these sophisticated technologies would further distance livestock farming from its traditional rural simplicity, and from the natural mating of animals to produce offspring. It would place technical demands upon the operators which favour large-scale company farms capable of supporting a range of highly trained specialists. Longworth identifies new techniques of tissue and cell culture as having 'the potential for enormous advances in crop improvement in the next couple of decades'.

Those techniques 'can both increase the genetic diversity and greatly increase selection efficiency', and they permit innumerable plants to be reproduced asexually from single cells or small pieces of tissue. The particular technique known as 'callus culture' has been used for many years to clone highly valued horticultural plants such as orchids, and cloning of cuttings is widely practised for tree crops such as tea and palm oil as well as by millions of gardeners for garden plants.

According to Longworth 'cell culture' of single cells has unexpectedly, and so far inexplicably, resulted in plants with different properties bcing regenerated from the same clump of parent tissue. Sugar cane, maize and potato plants regenerated in this way have been found which are resistant to important pathogens.

This application of cell culture with the capacity to generate vast numbers of seedlings rapidly has the potential to simplify greatly the hitherto labourious procedures of plant breeding and selection. However, in terms of current and immediate commercial importance, it is through micro-organisms that biotechnology has its greatest impact.

Microbial fermentation processes have been of great commercial significance for centuries, for example in bread, wine and cheese production, and there is the prospect of considerable development.

Two recent examples of new processes indicate the sorts of impact that such developments can have. In the late 1960s genetically engineered bacteria were developed which were able to digest corn starch to produce high-fructose corn syrup (HFCS) and which left as a residue corn gluten, which is now an important protein feed for livestock. HFCS has made considerable inroads in the United States and in Japan.

It has largely replaced sugar as a sweetener in Coca Cola as well as many other food and drink products. This has helped depress sugar prices to Third World growers.

In the European Community steps have been taken to prevent HFCS and other new powerful sweeteners from being produced and from undermining the market for domestically grown beet sugar, which is supported by the Common Agricultural Policy. A second important application of microbial fermentation has been the production of ethanol from sugar cane in Brazil and from maize in the USA.

The commercial viability of these processes is critically dependent upon the price of oil as the main non-renewable source of fuel, and, to date, massive subsidies have been required to maintain the Brazilian and USA ethanol programmes. Longworth states, however, that a new biotechnology, Sucrotech, is being patented, which will not only reduce the cost of producing ethanol from sugar cane, but will simultaneously produce fructose at a cost which will be competitive with HFCS and may thereby reclaim part of the sweetener market for sugar cane.

That such possibilities are in prospect indicates how volatile the future might be; biotechnology has tilted the competitive balance from sugar cane to maize, causing economic pressure and even disruption to sugar-cane-dependent economies and may in future switch it back again.

In the long run the capacity to ferment fuels microbially from renewable agricultural feedstocks points to an important long-term reorientation of agriculture if non-renewable oil becomes uncompetitively expensive as a fuel for cars and feedstock for certain chemicals.

It suggests that eventually there will be an increased emphasis on agricultural production of industrial feedstocks at the same time as continually increasing food output will be required to feed the expanding world population. All this will require considerable increases in agricultural productivity, to which biotechnology will increasingly contribute, but it will at the same time impose great strains on the natural environment and the structure of agriculture.

PRIVATE-SECTOR CONTROL OF BIOTECHNOLOGY DEVELOPMENT

It has been the role of the public sector to undertake research and development of agricultural technology as a public service to firms which might profit from translating that R & D into a commercial product or process, to farmers who might profit from adoption of the technology, and perhaps most importantly to consumers at home and abroad who benefited from the lower prices resulting from greater abundance.

This was particularly true of the phase of change dominated by improvements in plant and animal breeding, the 'old' biotechnology. While the public sector had a role in basic research for chemical and mechanical technologies, the benefits of research expenditure in these areas were easier to capture by private companies investing in R & D, so that increasingly the public sector has taken a smaller role in these areas although maintaining a strong regulatory role in regard to agricultural chemicals in particular.

Where it is impossible to prevent others from escaping payment for the research costs, either because the product is easily copied (seeds which can be regenerated by farmers or other firms) or because proposals for new methods can be readily implemented, private firms are understandably unwilling to invest. Machines, insecticides, fungicides and other manufactured inputs do lend themselves more readily to private exploitation.

Nevertheless the returns to R & D in these products do depend upon the degree of difficulty potential competitors would have in copying the product. In some cases there are inherent technical difficulties in copying the process, or it would be prohibitively expensive, but in other cases it is the ability to obtain patent protection which creates legal barriers to potential competitors' ability to become 'free-riders' and which protects incentives to private investment in R & D.

Traditionally, however, it has been impossible for plant and animal breeders to obtain patent rights for their products, which is one reason why public-sector R & D has remained so important in this area.

For, as under the European Patent Convention (EPC) of 1973, it has been judged impossible for plant varieties to satisfy one of the key criteria to qualify for a patent, namely proof of 'an inventive step'. The application of standard breeding practices to generate new varieties by crossing existing plants or animal strains has not been deemed to be invention.

Thus under the EPC one set of exclusions from patentability is 'Plant or animal varieties or essentially biological processes for microbiological processes or the products thereof.'

In the absence of patent rights plant-breeding firms in particular have worked hard to obtain other means of protection and royalties for their products. The history of the development of Plant Breeders' Rights (PER) is presented by Mooney, and is of particular interest because of the concerns he expresses about the consequences of allowing the basic genetic stock, which underpins agriculture and hence our whole society, from becoming private rather than public property.

While Mooney's concerns are expressed in relation to 'old' biotechnology varieties of plants, they are of particular importance with respect to the products of the 'new' biotechnology.

For bio-engineered products are capable of meeting the inventive-step criterion of patentability, and both micro-organisms and transgenic animals have now been patented in the USA.

Before examining developments in the seed industry it is worth touching upon the implications of changes in research policy which stress increasing reliance upon private R & D to develop and exploit the new biotechnology and other technologies.

In the UK government has decided that it should withdraw from funding what it terms 'near-market research', that is, research beyond the basic phase and which is preparation for commercial exploitation.

This policy is based in part upon the arguments that industry should be investing more heavily in R & D, and that research strategies at the near-market stage should be driven by assessments of likely commercial success which can best be made by the firms involved, and will therefore lead to greater efficiency in allocating research funds.

This has led to the closure of a number of government-financed agricultural research institutes, the scaling down of others and the sale of the National Seeds

Organization by auction to Unilever; Unilever won against competition with BP and other major public companies. This deliberate attempt to switch an increasing proportion of agricultural R & D expenditure from the public to the private sector carries with it a number of risks.

In the first place there is a controversy about whether there is under-investment in R & D so that the returns to extra expenditure are high, or whether the converse is the case. If there is under-investment, then withdrawal of public support for applied research will exacerbate this, as the private sector will only undertake R & D expenditure on those products and processes from which exclusive benefits can be captured by the investor.

Thus in relation to biotechnology in the USA, where again public support for applied biotechnology research is relatively small, Stallman and Schmid (1987) argue that there will be emphasis on technologies which are applied in factory conditions where secrecy and control can be maintained. For technologies which cannot be confined to factories, such as seeds, Stallman and Schmid state that 'Firms are also considering mechanisms which "scramble" the genome of a plant in the second generation', in order to prevent farmers from reproducing seed with the enhanced, engineered characteristics.

Clearly such actions are designed to frustrate the maximum spread of benefits from the technology and to maximize private profit for companies investing in research. Another facet of commercially orientated biotechnology research is that it will aim at the most important crops and livestock products, those produced by the largest farming units, and those which are most heavily subsidized.

In the latter case this will worsen the budgetary problems of adjusting agricultural policy in OECD countries. The probable neglect of minor crops, difficult habitats and small farmers means that the public sector will have a defined role in agricultural technology research, but one in which it will be relegated to the second division and where it is unlikely to prove successful in terms of the commercial yardstick of rates of return which is increasingly emphasized by public-research policy.

DEVELOPMENTS IN THE SEED INDUSTRY

Many major food crops upon which we depend originated and were first cultivated in Third World countries rather than in the industrialized countries where they are most productively exploited today and in which they have been improved by traditional plant breeding.

The potato originated in South America, wheat in Ethiopia and the Near East, important maize varieties in South America, and rice in Asia. Moreover the so-called Vavilov centres, which contain the greatest variety of living species and which have been the source of much important gene material, are mainly in Third World countries. The wealth of the industrial countries owes much to their past ability to exploit agriculturally crops and animals originating in other countries.

In the colonial era the success of British scientists in smuggling rubber plants from Brazil to Sri Lanka, Singapore and Malaysia and the comparable more complex route by which coffee was introduced to Latin America from Ethiopia are instances of colonial powers obtaining significant wealth by exploiting plant material from the Third World. In one way or another the process by which institutions and firms in developed countries have continued to collect new plant and animal varieties from less-developed countries has continued.

But what is still a major source of friction is the procedures by which legislative protection is developed to give companies in the richer countries commercial rights to exclusive exploitation of varieties developed from plants freely collected from other countries, and to extract high rates of profit from the sale of those varieties.

Most extreme is the situation whereby under the USA Plant Patent Act of 1930, which covered asexually propagated plants such as certain fruits, flowers, ornamental shrubs and trees, it has been possible for plants discovered and smuggled out of other countries to be given patent protection in the USA.

The key to this is that the Act accepted the unfeasibility of requiring proof of an 'inventive-step' for the granting of a patent, and employed the criterion that as far as can be determined the plant variety is new. Since some novel variety found in the rain forests of South America may satisfy this criterion, and the source of the plant can be concealed from the authorities, the theft and smuggling of plants can be said to have been encouraged.

Seed companies in developed countries have successfully pressed their case for protection, and have progressively extended Plant Breeders' Rights (PBR) through both national legislation and the Union for the Protection of New Varieties of Plant (UPOV), which is an international agreement among signatory countries to honour a system of varietal rights. In the case of food crops, in the absence of patentability, PBR operates by registering distinct varieties. Pressure has been exerted upon less-developed countries to become signatories of UPOV and thus acknowledge the rights of breeders, most of whom are from developed countries.

However, the majority have stalled or rejected membership, and have been fundamentally unhappy about accepting the principle of creating monopoly rights for others over plant material which in some cases originated in their own country.

The whole issue of intellectual property rights has been raised to a prominent political level by the USA's insistence that it be included as one of the thirteen areas for negotiation in the Uruguay Round of multilateral bargaining of the General Agreement on Trade and Tariffs. While the issue is a much broader one than that of plant and animal breeders' rights, it reflects a basic political conflict between the richer countries, determined to obtain secure returns on R & D expenditures by companies, and LDCs struggling to catch up and anxious to avoid having to pay royalties on products protected by PBR and patents.

This conflict has broader moral and ethical dimensions when it relates to the issue of charging royalties for the seeds of food crops to countries where malnutrition is widespread.

Against the previous argument has to be balanced the need of those incurring R&D expenditure to capture enough of the returns to provide an incentive for R & D. At least that is certainly the case where R & D is to be undertaken by the private sector, which is politically the increasingly preferred solution in the UK, the USA and elsewhere.

This, however, leads to another area of concern, which is that a small number of large multinational chemical and pharmaceutical firms have increasingly come to control the seed industry.

Companies such as Shell, Sandoz, Dekallb-Pfizer, Ciba-Geigy and BP have become owners of most of what were small independent seed companies. This linking between chemicals and seeds within firms has a strong commercial logic. It facilitates the marketing of packages consisting of seed varieties plus the appropri-ate fertilizers, insecticides, etc.

It also allows the companies concerned to integrate their plant breeding, biotechnology and chemical research. One particular synergy for these companies will be to bioengineer varieties of major food crops to resist their own herbicides and weedkillers.

This would permit chemical weed control to be extended to large-scale crops where this is not now possible, and is just one way in which the direction of biotechnolo-gical research may be influenced (biased) by the combination of forces giving a small number of companies extensive control over the basic means of agricultural production.

There are many other issues which could be touched on regarding corporate influence over the development of agricultural biotechnology. It is, however, understandable that the spokesmen for LDC countries, which are profoundly concerned about their dependence upon western technology and exercised by what is seen as colonial and post-colonial exploitation by the West, should be alarmed by the prospect of new rights being created over seeds and other biotechnological products.

Certainly there are widespread fears that they will only gain access to the fruits of biotechnology on unfavourable terms and that their relative dependence on and subservience to western industry will be increased so that benefits to them will be small. There are many western critics who would agree with this, and that the system by which agricultural biotechnology is delivered will favour the 'haves' rather than the 'have-nots'.

US FEDERAL REGULATIONS FOR AGRICULTURAL BIOTECHNOLOGY

Regulatory systems for the products of agricultural biotechnology have been in existence since the mid- to late 1980s within the United States. The regulatory approach to the safety evaluation of plants developed using rDNA technology has evolved in the best interests of research scientists, industry and the general public.

The agricultural products of rDNA technology, such as GM foods and crops, may require approvals from up to three regulatory agencies; the US FDA, the USDA and the US EPA, depending upon the characteristics exhibited by the GM plant, its proposed use and introduced traits.

The same standards of safety are applied to all products regardless of the technology used in their development. The US FDA is responsible for ensuring the human safety of all new foods and food components, including products developed using rDNA technology, under the Federal Food, Drug, and Cosmetic Act (FFDCA).

The USDA evaluates the potential of a GM plant to become a plant pest following its environmental introduction under the Federal Plant Pest Act (FPPA). The US EPA evaluates pesticides, including plant systems modified to express pesticides (*e.g.* insect-protected or virus resistance), under the Federal Insecticide, Fungicide, and Rodenticide Act (FIFRA).

As a result, the expression of an insecticidal protein in a food crop would undergo review by the USDA, US EPA and US FDA; a GM food crop exhibiting a modified oil content would be evaluated by the USDA and US FDA; and a non-food horticultural plant developed using rDNA technology for any other purpose (*e.g.* flower colour) would be subject to review by the USDA alone.

In most instances, obtaining all necessary approvals for the commercialization of an agricultural crop developed using rDNA technology takes a decade or more.

However, the exact amount of time required will depend on the need to confirm performance, to evaluate characteristics of the food, environmental effects, and to produce the required amount of seed before the product can be distributed and commercially grown by farmers.

Up to five years of field trials (5-10 generations of plants) are required for the developer of a new plant variety to collect sufficient data to meet the reporting requirements of the USDA. An additional five months to two years may be required for the US FDA, USDA and/or US EPA to complete all necessary product consultations, reviews and approvals. Approval for the first commercial planting of a GM food crop was not issued until 1995. Since 1995, more than 40 new agricultural crops developed using rDNA technology have received approval for commercial planting within the United States.

In 1999, approximately 72% of the total 39.9 million hectares (more than 98 million acres) of GM crops grown worldwide were planted in the United States.

Herbicide-tolerant soybeans (54%), Bt corn (19%) and herbicide tolerant canola (9%) accounted for approximately 82% of the GM plants cultivated. With an increasing number of agricultural biotechnology products reaching later stages of commercial development, it is anticipated that the overall area planted with GM crops will continue to rise.

The regulatory approach to evaluating the human health and environmental safety of GM crops within the United States is best described as a science-based, case-by-case assessment of hazards and risks.

This approach has provided the flexibility required to reduce the regulatory burden placed on products that have been determined to be of low risk or concern. All agencies involved in the regulation of plants developed using rDNA continue to implement and develop policies based on recommendations made within the Coordinated Framework.

In the future, the US FDA, USDA and the US EPA will be dedicating additional resources towards communicating how GM food and food components are regulated within the United States, and how these regulations function to be protective of both human health and the environment.

LIMITATIONS IN THE IMPACT ASSESSMENT OF BIOTECHNOLOGY

Two basic positions have been advocated in the socioeconomic impact assessment of biotechnology in Third World countries. On the one hand, there are those who see it as a panacea that will resolve all the food problems of the world; most typically the biotechnology companies themselves represent this position, *e.g.*, Monsanto's advertisements (Kleinman and Kloppenburg 1991).

On the other hand, there are those who posit a rather pessimistic, sometimes apocalyptic, scenario, seeing biotechnology as a First-World phenomenon that will tend to enhance two processes in agriculture: the appropriation by industry of increases in agricultural productivity and the tendency for industry to substitute agricultural for industrial products, *e.g.*, high fructose corn syrup for sugar. In this view, there are few hopes that Third World countries will access higher levels of economic development.

One example of the optimistic outlook, although not as extreme as that of the companies, is in the work of Robert Kalter, who did some of the first studies of the potential impact of bovine growth hormone in the U.S. on the basis of economic modelling. Another example of the optimistic outlook is represented by Cassio Luiselli Fernández (1987) for the case of Latin America. Even though the author presents both the opportunities and the threats offered by biotechnology, the final implication of his essay is that Latin American countries, and Mexico in particular, have a broad margin of action, if they only follow certain policy prescriptions.

There are at least three main areas in which one may identify an unfounded optimism: the assessment of scientific and industrial resources for biotechnology in developing countries, the assumed economic scale neutrality of new biotechnology products, and the possibilities and disposition of the various developing countries to establish cooperation efforts among them. The rest of this section deals with each of these points.

SCIENTIFIC AND INDUSTRIAL RESOURCES FOR BIOTECHNOLOGY

As far as scientific and technological production in Latin America, university–industry research relations are almost nonexistent. Universities in Latin America are more oriented to learning and teaching than to doing research. One might say that the only practical relation between universities and industry is that of supplying the latter with a qualified labour force for professional and operative work.

For the sake of illustration, let us look briefly at the case of Mexico, where university research is highly concentrated. The National Autonomous University of Mexico (UNAM) alone concentrates 50% of all research funds in the country. Nevertheless, UNAM's research budget represented only 22% of its global budget for 1988. The rest represented wages and other operating expenses (Ortega 1988).

The Mexican dilemma is that its science and technology system produces so little technology that most of it has to be imported. Still, there are four categories of university–industry links: training of professionals, creation of basic and applied knowledge, provision of technical services, and production of technology.

Part of the explanation for the weakness of university–industry links in Mexico is the near absence of a commercial criterion in research. Thus, the prevalent criterion tends to be defined in epistemological terms, which may be crucial for science but irrelevant for technology. For technology development, what matters is economic convenience, not epistemological relevance.

There are, however, some institutions doing cutting-edge biotechnology research in Mexico (Quintero 1985). In total, there are about 25 centres with biotechnology projects broadly defined. If we narrow the definition of biotechnology to the newer techniques, such as those based on recombinant DNA, the generation of hybridomas, cell and tissue culture, and protoplast fusion, then the number of institutions is reduced to only seven. The rest are involved in traditional biotechnology research. Of those seven, three belong to the National Autonomous University of Mexico, two to the National Polytechnical Institute, one to the Autonomous Metropolitan University and another one to Chapingo Autonomous University.

All of these universities are located around Mexico City, although three of the top research centres involved are decentralized in neighbouring cities (one in Irapuato and two in Cuernavaca).

The pioneering institution in promoting industry links has been the National Autonomous University of Mexico (UNAM), reputed as the largest university in Latin America. It created a "Centre for Technological Innovation" in 1984 to market UNAM's science and technology.

Although the explicit objective of the new centre is to market UNAM's science and technology, a former director declared to Expansión, the Mexican version of Fortune magazine, that they wish to preserve their identity as an academic institution.

They are not after tailor-made research, therefore, for the specific needs of companies. Rather, companies should find out what UNAM's researchers are doing that might be of interest to them. The director of one of the firms that has supported UNAM's research stressed that business firms should take advantage of Mexican technology, because it is cheaper and should allow them to compete better in international trade.

In sum, Mexico does have several excellent research groups working in the frontiers of biotechnology. But the number of researchers is very small in comparison to those in advanced capitalist countries, both in universities and in firms. In developing countries, there tends to be a very weak if not totally absent link between universities and industry.

With such great differences in the quantity of scientists and the differing patterns of university–industry linkages, it is hard to be very optimistic about the possibilities for developing countries to compete successfully in such a dynamic field as biotechnology on the basis of their own resources.

It is necessary, therefore, to undertake a more realistic analysis regarding the actual scientific potential of developing countries and what public policies it would take to translate research into products that might solve their specific socioeconomic and environmental needs. Unless the proper policy incentives exist, it is unlikely that these needs will be taken into account by the private firms doing biotechnology research, or by scientists from advanced countries who have other concerns as their central motivations.

ECONOMIC SCALE AND BIOTECHNOLOGY

Contrary to what optimistic analysts of biotechnology hold, it seems that the main tendencies indicate a bias favouring the large scale, rather than scale neutrality. That is to say, far from favouring small-scale farm operations, the new biotechnologies will tend to reinforce the bias introduced by modern agricultural technologies by requiring strong capital investments. With the new technologies then, the agricultures of developing countries are likely to require a smaller workforce, even if we assume that many of the products of biotechnology may be adopted by those countries.

The alternative might be that, if new technologies are not adopted, and given the global scope of today's world market, agriculture in developing countries may cease to be a viable economic activity altogether.

This would be a major paradox, for about 60% of the population in developing countries depends on agriculture. The following examples of new biotechnology products clearly reflect an economic bias towards large-scale production units. These are the implications of the "technological paradigm" represented by modern agriculture.

The generation and sale of cattle and sheep embryos is generally reserved for large and modern production units, which utilize the most advanced genetic technologies. Commercial cloning of livestock is becoming a reality. There are research groups in the U.S. that are developing cell fusion methods with the potential to create multiple, perhaps unlimited, copies of individual cow-and sheep-embryos.

The target in cloning is to produce genetically predictable animals, with superior milk-and meat-production capabilities. University of Wisconsin Professor Neal First and colleagues have successfully cloned cow embryos, but it will be years before they can determine the quality of the resulting breed. For what they clone is not the mother cow but the embryo itself, *i.e.*, an already fertilized egg.

One of the goals of researchers is to improve the cloning technique to develop the entire process in test tubes and eliminate the need for embryos from the cows. One implication of this type of research is that, in the future, it will be possible to produce "elite" animals for meat and milk production. This genetic improvement of animals, combined with the use of growth hormones, promises to revolutionize production in these fields, and they will probably displace large numbers of farmers, given the large-scale bias. The dictum for this type of modern operation will continue to be, perhaps in an exacerbated manner, "get big, or get out of farming."

Socially, this trend points to a future agriculture that will require not much more than a percent or two of a country's labour force. The main concern will be with how to employ productively, with adequate incomes, those people who are displaced from agriculture. In the U.S., this stage has already been reached: only 2% of the labour force is directly employed in farming.

Another factor that reinforces the previous interpretation regarding the economic bias towards the larger scale is that new agriculture will probably combine two or more products of the current technological revolution, including computers, for the integrated management of agricultural enterprises. It is very likely that a systems approach to management will be reinforced. For instance, and continuing with the example of livestock production, the most sophisticated dairy operations may use embryo transfer technologies to improve the herd, while using computers to maintain exact yield records per cow and to monitor their feed requirements.

The result will be that only large-scale operations will be able to make the heavy capital investments required for such integrated systems for managing resources. All of this reinforces the economic-scale tendency to discriminate against small-scale producers and accelerates the process of "depeasantization without full proletarianization". The dilemma involved in this process is that, although farming expels many of its workers because of increased capital-intensity, the larger economy cannot absorb all those redundant workers. This is the source of the swelling "informal sector" in Latin American economies.

PRODUCTION OF TRANSGENIC PLANTS

To produce transgenic plants, the crucial steps are the introduction of the cloned genetic material into the plant cell nuclear & the facilitated integration of the cloned gene into the plant chromosome.

It is interesting that the best method for doing so uses a system that already exists in nature, the system by which the plant pathogenic bacterium Agro-bacterium tumefaciens injects a portion of its plasmid DNA into plants & inserts it into the plant become.

The Agro-bacterium system is an excellent system for introducing foreign genes into the chromosomes of plants. The transfer into intact plant cells occur at high frequency, the T-DNA is usually integrated into the plant chromosomes at high frequencies without undergoing structural alterations, and the cells that have received the T-DNA can be selected easily by using antibiotic resistance such as the neomycin resistance maricer.

Finally the transgenic plants produced in this manner are quiet stable for at least several generations. Consequently almost all existing transgenic plants with potentially desirable traits have been obtained by this method.

HERBICIDE RESISTANCE PLANTS

The advantages of making crop plants resistant to herbicides are obvious. Although there is a fear that such resistant may eventually increase the use of herbiude chemicals, there are also reasons to except that these transgenic plants will promote the use of sager, more biodegradable herbicides, perhaps in smaller amounts one example involves the herbicide glyphosate which inhibits 5-enol-pyruvylshikimate 3-phophate synthase-an enzyme involved in the biosynthesis of aromatic amino acids. This enzyme has been purified from crop plants &

sequenced and DNA Probes corresponding to its amino acid sequence have been synthesized. These probes were used to isolate CDNA for the enzyme from the CDNA library of a plant cell line known to overproduce 5-enol-pyruvylshikimate 3-phophate synthase.

The CDNA was then cloned behind the strong CaMv promoter, and the promoter gene complex was introduced into plant cells via a disarmed Ti plasmid vector. The transgenic plants produce a much higher level of the target enzyme & thus are significantly more resistant to glyphosate. These results are encouraging because glyphosate has very low toxicity to animals is rapidly degraded in soil.

INSECT RESISTANT PLANTS

Bacitlus thuringiensis is used in the biological control of caterpillars because its sporulating cells contain toxic proteins. The gene for one toxic protein was cloned behind promoters that are effectively expressed in plants & was introduced into plant cells via a Ti plasmid vector, thus producing plants toxic to caterpillar.

The major problem with this approach has been the low level of expression of the toxin protein in plants, which is presumably related to the fact that the gene come from a bacterium,. Nevertheless, the method has produced tomato & tobacco plants that proved quiet resistant to caterpillars in field tests. Cotton however, is often attacked by insects that are more resistant to the B. thuringiensis toxin and the low level of expression of this toxin in transgenic cotton plants provided the plants with little if any protection.

Recently, the coding sequence of the toxin was altered extensively to replace codons that are rarely used in plants as well as to preclude the formation of a strong secondary structure in MRNA. When cotton plants were provided with this modified gene, their production of the bacterial toxin increased 100-fold and they showed impressive resistant to common bepidopteran insects than damage unmodified plants.

Some seeds contain high concentrations of protease inhibitors, which are thought to interfere with the digestive process in insects. The cloning of cowpea trypsin inhibitor genes, for example and their transfer to tobacco plants have resulted in good resistance to a wide variety of leaf-eating insects.

VIRAL RESISTANT PLANTS

The method most frequency used for producing virus-resistant plants sprang from the observation that, of ten times, plants infected by nearly avirulent virus are thereafter resistant to super infection by a related, highly virulent one. Thus, deliberate infection with avirulent virus strain has been used to produce protection in crop plants. The method is not totally safe, however, because the avirulent strains many mutate to produce strains that are significantly pathogenic. Most plant viruses are positive strand RNA viruses covered by coat protein sub units. When the virus enters an injured plant cell the replication process begins, starting with the progressive uncoating of the virus from the 5-end of the RNA.

Tranagenic plants, whose genomes contain introduced tobacco mosalc virus (TMV) coat protein genes and which continuosly synthesize the coat proteins, show resistance to virus infection. The phenotype of these plants is consistent with the idea that resistance is a result of interference with the uncoating of the virus particle. Although the plant cells are resistant to infect TMV particles, they remain sensitive to the TMV RNA or to partially uncoated TMV particles.

The coat protein gene is usually put behind a strong promoter such as the 35S promoter of CaMV & is introduced into a plant's genome via the Agro-bacterium Ti system. Transgenic plants showing significant resistance to TMV alfalfa mosaic virus, cucumber mosaic virus, tobacco streak virus & tobacco rattle virus have already been produced in this manner. In a recent experiment two virus & potato virus Y, were introduced simultaneously into a commercially important potato cultivar, and one of the resulting trangenic plants provide quiet resistant to both viruses under field test conditions.

APPLICATION OF CELL BIOTECHNOLOGY

Cell biotechnology or plant tissue culture is an enabling technology from which many novel tools have been developed to assist plant breeders. These tools can be used to increased the efficiency of the breeding process, to improve the accessibility of existing germplasm and to create new variation for crop improvement.

They include micropropagation, another culture, in vitro selection, embryo rescue, somaclonal variation, somatic bybridization and transformation. Of these somaclonal variation occupies a somewhat unique position, because it has both the advantages and disadvantages of tissue culture systems.

Somaclonal variation was first defined as such by Larkin and Scowcroft who reviewed the subject in 1981 and were among several authors at that time to draw attention to its potential use for crop improvement.

An extensive number of reports soon followed in a whole range of species, indicating that somaclonal variation was widespread, and therefore, ostensibly accessible to all plant breeders (Karp, 1991).

The Utilization of new genetic variability has become one of the major objectives of tissue culture, The assembly of genetic variability is vital for improvement of crop plants. Somaclones for the resistance of downy mildew. Fiji disease and eye spot disease were identifies. Application of somaclonal variation in crop improvement can produce increased yield and resistance to biotic as well as abiotic a stresses. Some of the crops where somaclones were produced are rice, wheat, maize, tomato, geranium, sweet pototo, sugarcane, rice, celery and brown mustard. Our country research efforts have developed a variety called 'Pusa Jaikisan' in brown mustard evolved by somaclonal variation.

Table. List of varieties developed from somaclonal variation.

Crop	Country and Institution	New trait
Geranium Univ USA	Deptt. of Hort, Purdue Vigour and attractive flowers	
Sweet potato	North Carolina Res, Services, quality of roots	USA Colour, shape and backing
Sugarcane	Sugar Research Centre, Fiji	Yield and disease resistance
Maize	Motecular Genetics, USA	Trytophan content
Tomato	DNA Pl. Tech. of New Jersey, USA	Dry matter content, Disease resistance
Rice Japan Unv of Agri. Sci., Godollo, Hungary	Plant Research Institute, Yield Disease resistance	
Celery New Jersey, USA	DNA Pl. Tech. of Processing, Yield & efficiency	
Brown mustard	ICAR, New Delhi	Yield.

Micropropagation is another technique of great significance to agriculture which is based on in vitro culture of plant cell. Each and every cell of a plant has the potency to be regenerated into complete plant by tissue culture techniques and this property of the cell is known as totipotency.

Using this technique hundreds and thousands of seedlings can be generated in a very short time, starting from a limited number of explants. This methods of multiplication has two major advantages.

First is the multiplication of new elite cultivars. For example, after its release, a new sugarcane cultivar may take 5-10 years before sufficient planting material may be generated for wide coverage in the farmers field. But using micropropagation techniques a comparable level of multicipation may be achieved within 2 years.

The second advantage is that we get disease free planting material because of the use of aseptic conditions during tissue culture. There is also reports of rejuvenation and increased vigous following tissue culture. This can also help in rescue of important materials which are on the verge of extinction due to disease attack. For example, a rare variety of banana used for temple offering in Karnataka has been rescued through tissue culture techniques.

Micropropogation of fruit and ornamental plants provides a useful avenue for the employment generation for rural youths at cottage level industries. In species like Daffodies and gladioli, microporpagation techniques are being used to speed up the release of new varieties. In strawberry, millions of plants can be produced from a single mother plant in one year.

Micropropagation work undertaken at our Defence Agricultural Research Laboratory. Pithograh was resulted into successful multiplication of strawberry, petunia, carnation and tomato hybrid.

In case of strawberry a well adapted variety of hill vi., Jyolikit Selection was taken for micropropagation studies with objective to large scale production of disease free propagules since this fruit plant is highly susceptible to number of fungal, bacterial and viral diseases. The result showed that internodal segments are most suitable explant material for micropropagation under in Vitro condition.

MS formulated with 1ppm of BAP and 0.1ppm of GA3 was found optimal protocal for maximizing shoot proliferation in explant where MS media modified with 1 ppm 1AA was most suitable for early rooting in shootlets. Peat moss was found to be most suitable potting culture for hardening of tissue culture raised plants of strawberry var. Jyolikot Selection.

Micropropagation studies in Petunia Hybrida were undertaken in order to raise the plant seedlings in mass.

The result showed that optimum number of shootlets could be obtained after culturing the leaf explant of size 5mm X 2.5mm in MS media supplemented with 1ppm of BAP + 0.5 ppm of IAA with an inclubation period of 34 days. Protocol standardisation revealed that MS medium supplemented with 1 ppm of IBA was suitable for an early and healthy rooting in tissue culture raised shootlets. Hardening process was most successful with high relative humidity (90-95%), Temperature 23 + 10 C and potting mixture of sand and soil in equal proportion.

Work on micropropagation of carnation was resulted into standardization of an vitro culture media for shoot and root proliferation. Protocol found most promising for shoot proliferation was MS medium supplemented with 1 ppm of BAp AND 0.1 ppm of IAA where MS medium supplemented with 1 ppm of NAA was identified to most suitable for rooting in shootlets.

Micropropagation studies were also undertaken in hybrid tomato to generate large scale seedings to fulfil the need of army personnel as well as local growers with pure and healthy planting material.

Based on the observation it was found that tissue culture medium comprised of MS,0.5 ppm kinetin, 0.5 ppm IAA and 0.3 ppm GA3 was most suitable for shoot growth and MS medium supplemented with ppm of IAA or MS with 1 ppm of IBA was most suitable for early and healthy rooting in the explant material.

Tissue culture regenerated plantlets showed high survival under relative humidity varied from 85-95 per cent, temperature 25+20C and potting mixture sand and oak forest humus in equal ratio.

APPLICATION OF DNA TECHNOLOGY

Genetic engineering opens a totally new dimension for bioprospecting. The search for new genes and their application is the primary objective of the biotech industry. Gene technology now enable humans to integrate revolutionary new properties in to cultivated plants through inter-specific or inter-generic gene transfer which was not possible through classical approach of crop improvement.

Genetic engineering is the direct introduction into a plant of an isolated or modified single gene using transformation techniques. Now it is possible to transform almost all of the plants cultivated by man. Where this has not happened, it is more to do with the insignificance of the plant in agriculture or forestry rather than a reflection of a stubborn resistance to all attempts to transform it. A couple of years ago, cereals were regarded as being recalcitrant species. This has new changed and transformation has been reported as being successful for nearly all cereal species.

TRANSGENIC PLANTS

Transgenic is used to describe plants which have had DNA introduced into them by means other than by the transfer of DNA from a sperm cell to an egg. Development achieved through modern biotechnology in the form of transgenics have given new options of controlling insects, diseases and weed pests, while improving the overall integrity and consequences of agricultural practices.

Biotechnology combined with modern crop management techniques and the responsible use of pesticides gives the best tools available to ensure healthy, high yielding crops.

However, the benefit expected from the release of genetically modified organisms into environment are quite more which are seems to pay major role in bioremediation, environmental improvement, agriculture, food industry and health care. A number of transgenic plants in various crop species covering wide range of altered characteristics have been approved for commercial cultivation.

Resistant to Biotic Stresses

Damage to crops by biotic factors like insects-pests and weeds is a major limiting factor in agriculture economic in tropical and temperate regions of the world. Despite large scale investment in the chemical control of pests, they cause great economic loss by damaging or destroying the crops.

With a tremendous development in techniques to engineer crops genetically, we now have the ability to make broader use of natural insecticides. The choice of a Bt endotoxin, as the first insecticidal protein for introduction into plants, was based on the extensive knowledge gathered about this class of crystal protein since 1902 and such a strategy has been successfully achieved.

The application of glyphosate kills crop plants just as effectively as it kills weeds. The usefulness of glyphosate as a weed control agent in agriculture could be enhanced if resistance to herbicide could be selectively engineered into crop plants.

Transgenics have been developed by introducing the corresponding resistance genes from weeds and microbes into crop plants such as tobacco. In addition genes conferring resistance against herbicides, *viz.*, Sulphobylurea, Imidazolinones, Phosphinothricin, Asulam, Bromoxynil and 2,4-D have been engineered and transgenic plants developed.

Abiotic Resistance

A large area of our country is under stress conditions *i.e.* saling, alkaline water, cold and heat stress etc. transgenics produced from introducing mtl 1D gene from E. coli in tobacco and arabidopsis showed tolerance to high leave of NaCl. Thermoc tolerance was found in Arabidopsis plants engineered with hsf gene.

It has been reported that over-expression of sac B gene from Bacillus subtillis leads to high level of fructans in tobacco cells, and this is associated with increased drought tolerance whereas chemically synthesized antifreeze protein genelala 3 leads to improved freezing tolerance of tobacco plants.

Genetic engineering for improved tolerance to abiotic stresses is, therefore, the need of the hour as the existing cultivars in most cases are capable of giving much higher biomass that what we harvest today.

While it is true that the response of plants to these stresses is multigenic, the recent success achieved in genetic engineering against excess salt, high and low temperature as well as less or excess water by altering individual gene is noteworthy.

Product Quality

A group of scientists have reported the ethylene synthesis in transgenic tomato plants, using anti-sense to a gene that has been identified as ACC oxidase. The enzyme convert ACC to ethylene.

Ethylene production was inhibited by 97% with a signification reduction in over-ripening and shrivelling. Flavr Savr and Endless Summer in tomato, Freedom II in squash, high lauric acid reposed (Canola) and Round Up Ready soybean are some examples of the crops that already being commercially grown in developed countries.

Genetic engineering of metabolic pathways of fruits and vegetables has therefore given control over post-harvest process which not only leads to improvement of quality characteristics but also enhances processability, transportability and prologation of shelf like. In addition, engineered plants with change fatty acid composition of edible oil, reduced antinutritional components in many food crops, production of palm oil or animal derived industrial oil into plants like Brassica are some of the good examples of novel products that can be expected from biotechnology.

Significant progress has been achieved for production of biodegradable plastics derived from biomsass of genetically altered plants having gene from bacteria. With the advancement in DNA technology and understanding the structure and function of gene and its products more and more transgenics are coming out from various research programme all over the world.

However, boon of biotechnology has been confined to few developed countries and developing countries are still in vogue to harvest the benefits of modern technology.

More that 48 transgenic has picked up remarkably in last couple of years. During 1986-87, 25000 field trials were conducted with transgenics of more that 60 crops in 45 countries. Sixty per cent of these trials were in first 10 years and 40% in the last two years. Area under transgenics cultivation has also increased tremendously. During 1996, the transgenic crops covered an area of 2.8 million has which increased about 4 times during 1997 and 10 times during 1998.

Table. Area of transgenic crops planted (m ha).

Country	1997	1998
USA	8.1	20.5
Argentina	1.4	4.3
Canada	1.3	2.8
Australia	0.1	0.1
Mexico	<0.1	0.1
Spain	0.0	<0.1
France	0.0	<0.1
South Africa	0.0	<0.1
Total	11.0	27.8

A major challenge for agriculture in the 21st century in developing nations would be to speed up production process economically to fulfil the aspirations of huge populace, to achieve diversification and adding value to the primary produce so as to make agriculture enterprise farmers as well as environmental friendly. Biotechnology based advance technologies are expected to materialize many of our expectations in the coming millennium.

Viruses

To minimize the losses due to viral disease and to reduce the use of chemicals, the cross protection phenomenon of viral coat protein and movement proteins have been employed. This method has been used to develop transgenic plants which conferred significant levels of resistance in tobacco plants to TMV or to alfa mosaic virus (AMV) using the AMV coat protein gene.

In some viruses it has been found that certain sequences termed satellite RNAs act to reduce disease symptoms in infected plants and as such it is being utilized for developing transgenic with virus resistance. Antisense RNA which binds to sense RNA and artificial ribozyme molecules which cleave specific gene transcripts were found very powerful tools for the inhibition of viral disease development. Genetic engineering of antisense or ribozyme genes complementary to viral genes whose products are essential may prove one of the best way of viral disease control.

Insects

Transgenic crop plants that produce pesticidal proteins such as Bt toxin can offer the growers advantage of increased yields, numerous environmental benefits due to decreased use of conventional insecticidals. A transformed tobacco plants with the gene encoding the cowpea trypsin inhibito, in feeding experiments demonstrated that expression of the gene conferred resistance to a range of insect genera, including Heliothis, Spodoptera, Diabrotica and Tribolium the protein is not however toxic to humans.

Herbicides

The use of herbicidea has become an important feature of modern agriculture as a means of controlling weed species, however, it is economically feasible to produce specific herbicides for use with every individual crop species. A more generic approach would be to transform a range of crop plants with a gene which confers tolerance to a single non-selective herbicide.

AGRICULTURAL BIOTECHNOLOGY AND THE DEVELOPING WORLD

Biotechnology has the potential to play a large role in more rapidly advancing agricultural productivity in developing countries while protecting the environment for future generations, writes J.B. Penn, under secretary for farm and foreign agricultural services at the U.S. Department of Agriculture. Penn says biotechnology is simply another crop improvement tool in the long history of cultivation.

Agricultural biotechnology has been changing the face of agriculture since its commercial introduction in 1996 and the widespread adoption of bioengineered crops by farmers in the United States and other countries.

However, this technology is not without controversy and is causing political reverberations around the world. While it holds enormous promise for significantly increasing food production and relieving already strained land and water resources, it has become an emotional issue among some consumers and environmental groups. As the science continues to be developed, it clearly will present both opportunities and challenges to participants throughout the food chain.

BACKGROUND ON CONVENTIONAL PLANT BREEDING

Almost all plants can be considered "genetically modified." Genetic modification occurs when plants within a species simply produce offspring. The offspring is not exactly like either of the parents; it is a genetic combination of both. For centuries, plants have been cultivated and cross-bred by man to produce offspring with specific, desired traits.

For example, maize as we know it today barely resembles its ancestor, teosinte, or Zea mexicana, a tall grass that produces finger-length "ears" containing a single row of a few grains.

Maize produced today has been cultivated for many years to serve as a food crop, with far different traits than those of its predecessors. When varieties are cross-bred to produce a hybrid plant, millions of genes are combined in the process.

Scientists must select and continually cross-breed the plants, often over a period of several years, to obtain plants with the largest number of desired traits and the least number of undesirable traits.

AGRICULTURAL BIOTECHNOLOGY TODAY

While the focus of the first "generation" of biotech crops has been on the considerable economic benefits to farmers, more and more evidence is accumulating that significant food safety and environmental benefits are beginning to accrue.

Farmers have indicated their acceptance of biotech varieties by the unprecedented pace in which they have been adopted. According to the U. S. Department of Agriculture (USDA), in the United States approximately 80 per cent of soybeans, 38 per cent of maize and 70 per cent of cotton were planted to biotech varieties in 2003.

The United States is not alone in experiencing this evolution in agriculture. Adoption rates in other countries, such as Argentina, Canada and China, where biotech varieties are approved, have been similarly rapid.

According to the National Centre for Food and Agricultural Policy in Washington, D.C., U.S. farmers have realised the following benefits through the use of biotech varieties:

- *Roundup Ready Soybeans*: 28.7 million lbs. (13,018.3 metric tons)/year decrease in herbicide use; $1.1 billion/year savings in production costs.

- *Bt Cotton*: 1.9 million lbs. (861.8 metric tons)/year decrease in insecticide use; 185 million lbs. (83,916 metric tons)/year increase in cotton production.
- *Bt Maize Varieties*: Over 16 million lbs. (7,257.6 metric tons)/year decrease in insecticide use; 3.5 billion lbs. (1,587,600 metric tons)/ year increase in production volume.
- *Papaya*: Virus-resistant biotech papaya saved the Hawaiian papaya industry $17 million/year in 1998 from the devastating effects of ringspot virus. These results illustrate enormous decreases in pesticide use, with corresponding environmental enhancement, along with equally dramatic increases in production and savings in production costs. While biotech results vary by farm, the economic benefits obviously have been significant. These benefits are realised not only by farmers, but also by the environment and to consumers in general.
- The reduced reliance of biotech varieties on chemical inputs means less water pollution.
- Reduced chemical usage results in safer water supplies and higher quality drinking water as well as a better environment for wildlife.
- Higher yielding biotech crops can help ease the strain on land resources, reducing the need for expansion onto more fragile areas and thus allowing for greater conservation of natural habitats.
- Energy usage on biotech crops is lower because there are fewer passes through fields in applying chemicals. Less fuel use means less carbon entering the atmosphere as carbon dioxide (CO_2).
- Herbicide-resistant crops encourage the adoption of conservation tillage, especially no-till, which reduces erosion of topsoil.

CURRENT RESEARCH ON MODERN BIOTECHNOLOGY

Current research will lead to food crops that are resistant to environmental pressures such as drought, temperature extremes and salty soil. Scientists around the world are also investigating the "second generation" of biotech products - those with direct consumer benefits such as enhanced nutrition levels.

Many of us have heard of "golden rice," which contains higher levels of beta carotene–an important component in vitamin A production. Scientists in India are working to develop a biotech potato variety with higher levels of protein. Edible vaccines could also be produced by plants to provide low-cost, low-maintenance medicines. These are just a few of the numerous examples of cutting edge research that will further the changes we have already witnessed in the global food chain. The possibilities are enormous.

IMPLICATIONS FOR THE DEVELOPING WORLD

Global population projections suggest an additional 725 million mouths to feed in just 10 years. By 2020, this will grow to 1.2 billion more people to feed—equivalent to the populations of all Africa and South America combined. This expansion comes despite the fact that today some 800 million people - nearly one in seven—face chronic hunger. This is especially devastating to the world's children, where one in three is undernourished, and a child dies every five seconds due to hunger.

Biotechnology alone will not feed tomorrow's world. However, this far-reaching agricultural technology, in combination with political and economic reforms, can increase crop productivity by increasing yields and improving the nutritional content of crops in developing countries. It will also help provide lower-cost food to low-income consumers.

Bringing such benefits to developing countries would have far-reaching results, indeed. An annual increase of 3 to 4 per cent in African crop and livestock yields would almost triple per capita incomes while reducing the number of malnourished children 40 per cent.

Increased agricultural productivity will drive economic growth and expand opportunities to trade, bringing more and better jobs, better health care, and better education. Consumers in developing countries spend a high proportion of their disposable income on food, which could be reduced with a more efficient food system, thereby leaving more of their income for other products to enhance their quality of life.

The most critical areas in the world for bringing economic prosperity and stability are the developing countries. Agricultural productivity in these countries must advance more rapidly to meet growing food demand and raise incomes while protecting the environment for future generations. Biotechnology has the potential to play a large role in this achievement.

MODERN BIOTECHNOLOGY

Modern biotechnology is a tool that allows scientists to select a single gene for a desired trait, incorporate it into plant cells, and grow plants with the desired trait. In many ways it is simply a "high-tech" version of traditional plant breeding. This more efficient process prevents millions of genes from being crossed and possibly producing undesirable traits. Biotechnology is also different because it allows scientists to incorporate genes from other species—something that cannot be done via conventional plant breeding.

This makes biotechnology a very powerful and useful tool for plant breeders. Some people fear this tool because it is perceived as "unnatural." However, most people forget that the food crops we have today would not exist without man's intervention, whether through plant breeding, fertilizer application, delivery of irrigation water or use of modern tractors and equipment. Without cultivation by man over the years, we would still have teosinte instead of conventional maize. The same is true for wheat, tomatoes, potatoes, watermelon and any product on today's supermarket shelf. Thus, biotechnology is simply a modern, additional tool in the long history of plant cultivation and agriculture.

UNDERSTANDING BIOTECHNOLOGY IN AGRICULTURE

Bioengineering provides distinct advantages over traditional breeding technologies because the risk of introducing detrimental traits is likely to be reduced, says Deputy U.S. Food and Drug Administration Commissioner Lester Crawford. Crawford, a doctor of veterinary medicine by training, argues that there are no scientific reasons that a product should include a label indicating that it, or its ingredients, was produced using bioengineering.

He also outlines draft guidelines to strengthen controls that would prevent biotech products in field trials from inadvertently getting into food or feed.

Based on two decades of experience with bioengineered foods and overwhelming scientific data that these foods are safe to eat, we believe that biotechnology can offer a safe and important tool for both exporting and food deficit countries. This chapter describes some of the basic science behind biotechnology, the U.S. regulatory structure for ensuring safe foods and U.S. policy on the issue of labeling.

CROSS-BREEDING, HYBRIDIZATION AND BIOENGINEERING

Scientists have been improving plants by changing their genetic makeup since the late 1800s. Typically, this has been accomplished through cross-breeding and hybridization, in which two related plants are cross fertilized and the resulting offspring have characteristics of both parent plants. In the breeding process, however, many undesirable traits often can appear in addition to the desirable ones. Some of those undesirable traits can be eliminated through additional breeding, which is time consuming.

Breeders can then further select and reproduce the offspring that have the desired traits. Many of the foods that are already common in our diet are obtained from plant varieties that were developed using conventional genetic techniques of breeding and selection.

Hybrid maize, nectarines, which are genetically altered peaches, and tangelos, which are a genetic hybrid of a tangerine and grapefruit, are all examples of such breeding and selection. Today, by inserting one or more genes into a plant, scientists are able to produce a plant with new, advantageous characteristics.

The new gene splicing techniques are being used to achieve many of the same goals and improvements that plant breeders historically have sought through conventional methods.

They give scientists the ability to isolate genes and introduce new traits into foods without simultaneously introducing undesirable traits. This is an important improvement over traditional breeding. Because of the increased precision offered by the bioengineered methods, the risk of introducing detrimental traits is actually likely to be reduced.

FOOD SAFETY CONCERNS

The U.S. Food and Drug Administration (FDA) have found no evidence to indicate that either ordinary plant deoxyribonucleic acid (DNA) or the DNA inserted into plants using bioengineering presents food safety problems. Nor are the small amounts of the newly expressed proteins likely to change dramatically the safety profile of the plant.

If safety concerns should arise, however, they would most likely fall into one of three broad categories: allergens, toxins or anti-nutrients. FDA has extensive experience in evaluating the safety of such substances in food.

It is important to note that the kinds of food safety testing typically conducted by developers of a bioengineered food crop to ensure their foods meet all applicable requirements of the Food, Drug and Cosmetics Act (FD&C Act) address these potential concerns.

In the event that something unexpected does occur, this testing provides a way to detect such changes at the developmental stage and defer marketing until any concern is resolved.

As aforementioned, some of the food safety concerns that could arise include:

- *Allergens:* Foods normally contain many thousands of different proteins. While the majority of proteins do not cause allergic reactions, virtually all known human allergens are proteins. Since genetic engineering can introduce a new protein into a food plant, it is possible that this technique could introduce a previously unknown allergen into the food supply or could introduce a known allergen into a "new" food.
- *Toxins:* It is possible that a new protein, as introduced into a crop as a result of the genetic modification, could cause toxicity.
- *Anti-nutrients:* It is possible that the introduction of anti-nutrients, such as molecules like phytic acid, could reduce essential dietary minerals such as phosphorus. The use of genetic engineering techniques could also result in unintended alterations in the amounts of substances normally found in a food, such as a reduction of Vitamin C or an increase in the concentration of a naturally occurring toxicant in the plant food.

LEGAL AND REGULATORY ISSUES

One important component in ensuring food safety is the U.S. regulatory structure. The FDA regulates bioengineered plant food in conjunction with the United States Department of Agriculture (USDA) and the Environmental Protection Agency (EPA). FDA has authority under the FD&C Act to ensure the safety of all domestic and imported foods for man or animals in the United States market. The exceptions to this are meat, poultry and certain egg products, which are regulated by USDA.

The safety of animal drug residues in meat and poultry, however, is regulated by FDA. Pesticides, including those bioengineered into a food crop, are regulated primarily by EPA. USDA's Animal and Plant Health Inspection Service (APHIS) oversees the agricultural and environmental safety of planting and field testing bioengineered plants. Bioengineered foods and food ingredients must adhere to the same standards of safety under the FD&C Act that apply to their conventionally bred counterparts.

This means that these products must be as safe as the traditional foods in the market. FDA has the power to remove a food from the market or sanction those marketing the food if the food poses a risk to public health. It is important to note that the FD&C Act places a legal duty on developers to ensure that the foods they market to consumers are safe and comply with all legal requirements.

FOOD ADDITIVES

A substance that is intentionally added to food is a food additive, unless the substance is generally recognized as safe (GRAS) or is otherwise exempt, such as a pesticide whose safety is overseen by EPA. The FD&C Act requires premarket approval of any food additive regardless of the technique used to add it to food. Thus, substances introduced into food are either new food additives that require premarket approval by FDA, or GRAS and are therefore exempt from the requirement for premarket review.

Generally, foods such as fruits, vegetables and grains are not subject to premarket approval because they have been safely consumed over many years. Other than the food additive system, there are no premarket approval requirements for foods generally. Under FDA policy, a substance that would be a food additive if it were added during traditional food manufacturing is also treated as a food additive if it is introduced into food through bioengineering of a food crop. Our authority permits us to require premarket approval of any food additive and, thus, to require premarket approval of any substance intentionally introduced via bioengineering that is not generally recognized as safe.

Examples of substances intentionally introduced into food that would be reviewed as food additives include those that have unusual chemical functions, have unknown toxicity, or would be new major dietary components of the food.

For example, a novel sweetener bioengineered into food would likely require premarket approval. In our experience with bioengineered food to date, however, we have reviewed only one substance under the food additive provisions, an enzyme produced by an antibiotic resistance gene, and we granted it approval as a food additive. In general, substances intentionally added to or modified in food via biotechnology to date have been proteins and fats that are, with respect to safety, similar to other proteins and fats that are commonly and safely consumed in the diet and, thus, are presumptively GRAS. Therefore, they have not needed to go through the food additive approval process.

PRE-MARKET CONSULTATIONS

We are not aware of any bioengineered plant food that is subject to FDA's jurisdiction and is on the market that has not been evaluated by FDA through the current consultation process. Typically, the consultation begins early in the product development stage, before the product is ready for market. Company scientists and other officials meet with FDA scientists to describe the product they are developing.

The agency then advises the company on what tests would be appropriate for the company to assess the safety of the new food. After the studies are completed, the data and information on the safety and nutritional assessment are provided to FDA for review. FDA evaluates the information for all of the known hazards and also for potential unintended effects on plant composition and nutritional properties since plants may undergo changes other than those intended by the breeders. For example, FDA scientists are looking to assure that the newly expressed compounds are safe for food consumption and that there are no allergens new to the food, no increased levels of natural toxicants, and no reduction of important nutrients.

They are also looking to see whether the food has been changed in any substantive way such that the food would need to be specially labeled to reveal the nature of the change to consumers. If a plant developer used a gene from a source whose food is commonly allergenic, FDA would presume that the modified food might be allergenic.

The developer, however, is allowed the opportunity to demonstrate that such food would not cause allergic reactions in persons allergic to food from the source. Our experience has been that no bioengineered product has gone on the market until FDA's questions about the safety of the product have been answered.

LABELING

One of the most important issues confronting the biotechnology industry is that of labeling. Under the FD&C Act, a food is misbranded if its labeling is false or misleading in any particular way. FDA does not require labeling to indicate whether or not a food or food ingredient is a bioengineered product, just as it does not require labeling to indicate which conventional breeding technique was used in developing a food plant. However, if genetic modifications materially change the composition of a food product, these changes must be reflected in the food's labeling.

This would include its nutritional content (for example, more oleic acid or greater amino acid or lysine content) or requirements for storage, preparation or cooking, which might impact the food's safety characteristics or nutritional qualities. For example, one soybean variety was modified to alter the levels of oleic acid in the beans.

Because the oil from this soybean is significantly different from conventional soybean oil, we advised the company to adopt a new name for that oil, a name that reflects the intended change.

If a bioengineered food were to contain an allergen not previously found in that food and if FDA determined that labeling would be sufficient to enable the food to be safely marketed, FDA would require that the food be labeled to indicate the presence of the allergen.

FDA has received comments suggesting that foods developed through modern biotechnology should bear a label informing consumers that the food was produced using bioengineering.

We have given careful consideration to these comments. However, we do not have data or other information to form a basis for concluding that the fact that a food or its ingredients were produced using bioengineering constitutes information that must be disclosed as part of a bioengineered product's labeling.

Hence, we believe that we have neither a scientific nor a legal basis to require such labcling. We have developed, however, draft guidance for those who wish voluntarily to label either the presence or absence of bioengineered food in food products.

STRENGTHENING CONTROLS OVER FIELD TRIALS

In August 2002, President Bush's Office of Science and Technology Policy (OSTP) proposed strengthening controls over field trials to address the potential of material from field trials inadvertently getting into food or feed.

FDA's task is to publish draft guidance for comment on procedures to address the possible intermittent, low-level presence in food and feed of new non-pesticidal proteins from biotechnology-derived crops that are under development for food or feed use but have not gone through FDA's premarket consultation process.

Under this guidance, FDA would encourage sponsors, domestic and foreign, to submit protein safety information when field testing showed that there could be concerns that new non-pesticidal proteins produced in the field-tested plants might be found in food or feed. FDA's focus would be on proteins new to such plants because FDA believes that at the low levels expected from such material, any food or feed safety concerns would be limited to the potential that a new protein could cause an allergic reaction in some people or could be a toxin.

PHARMACEUTICAL CROPS

FDA has the authority and responsibility for regulating pharmaceuticals, whether they are manufactured in a traditional manufacturing plant or manufactured in crops in the field.

For crops in the field, however, there are additional issues to be addressed, including issues involving the parts of the plant that do not contain the pharmaceutical and the residual crop left over after a pharmaceutical is extracted.

In September 2002, FDA and USDA published Draft Guidance for Industry on the use of bioengineered plants or plant materials to produce biological products, including medical devices, new animal drugs, and veterinary biologics.

This draft guidance outlines the important scientific questions and information that should be addressed to FDA by those who are using bioengineered plants to produce medical or veterinary products. We are currently reviewing public comments on this guidance.

After 10 years of experience in this country, there is every reason to conclude that bioengineered foods are as safe as food produced through traditional breeding techniques. Both the U.S. General Accounting Office (GAO) and the National Academy of Sciences (NAS) have issued reports agreeing with this assessment.

We are confident that the foods developed using bioengineering that we have evaluated are as safe as their counterparts, and we will continue to follow the development of this technology to ensure that any new safety questions are also resolved prior to marketing.

COTTON

In 2000, 72% of U.S. cotton acreage was planted to insect and/or herbicide resistant varieties: 39% *Bt*, 61% herbicide resistant and 28% "stacked" varieties with both insect and herbicide resistance (USDA AMS).

Cotton is the fifth largest crop in the U.S., in terms of acreage in production, planted on over 15 million acres in 2000. In 1999, nearly 17 million bales of cotton were harvested, worth nearly $4 billion (USDA NASS 2000d). Cotton production is concentrated in the south and west.

Texas is the largest cotton producing state, accounting for 40% of total U.S. acreage and 30% of total production in 2000.

Georgia has recently expanded acreage to become the second largest cotton producing state, in terms of area in production, planting 9% of U.S. total acreage in 2000. California, North Carolina and Arizona each have nearly 1 million acres in cotton production.

BT COTTON

Bt cotton varieties were introduced in 1996, providing control of three major cotton insect pests: tobacco budworm, cotton bollworm and pink bollworm. Cotton bollworm and tobacco budworm are prevalent in the Southeast and Mid-South production areas, while pink bollworm infests cotton acreage primarily in western states such as Arizona and California.

These varieties offer an alternative to conventional insecticide spray programmes. Insecticides were used on 75% of the total cotton acreage before the introduction of *Bt* varieties (USDA NASS).

In 1995, the year before *Bt* varieties were introduced, it was estimated that 2.4 insecticide applications per year, on average were made to control bollworm/budworm across all cotton producing states, and that a 4% yield loss was incurred due to these two pests. Tobacco budworm infestations were particularly heavy in 1995, causing severe yield losses in some areas.

Table. USDA AMS Bt Cotton Adoption Estimates.

	Percent of planted acres				
	1996	**1997**	**1998**	**1999**	**2000**
Alabama	65	80	61	75	65
Arizona	11	64	57	56	57
Arkansas	15	14	14	22	60
California	0	2	5	8	6
Florida	42	79	80	66	75
Georgia	23	34	47	51	48
Louisiana	21	32	71	67	81
Mississippi	37	45	60	66	75
Missouri	0	0	0	1	5
New Mexico	9	13	38	28	36
North Carolina	3	2	4	37	41
Oklahoma	2	7	2	47	55
South Carolina	16	27	17	84	71
Tennessee	0	6	7	59	76
Texas	1	4	7	9	10
Virginia	0	3	1	16	41
US	12	18	23	32	39

The worst damage was sustained by Alabama growers, who on average experienced a 29% yield loss due to bollworm/budworm (Williams 1996). These losses were attributed to the ineffectiveness of pyrethroid insecticides against budworm, due to the development of resistant populations in some states.

Two surveys provide estimates of Bt cotton adoption. The USDA Agricultural Marketing Service estimates are based on informal surveys of cotton ginners, seed dealers, extension agents and others. In addition, the National Cotton Council sponsors an annual survey that provides estimates of adoption of Bt cotton varieties.

The Cotton Council survey is coordinated by entomologists in each of the states, collecting information derived from surveys of county agents, extension specialists, private consultants and research entomologists. The estimates of adoption of Bt cotton from USDA AMS and the Cotton Council. Adoption estimates for 2000 are not yet available from the Cotton Council.

Table. Cotton Council Bt Cotton Adoption Estimates.

	Percent of planted acres			
	1996	**1997**	**1998**	**1999**
Alabama	77	68	63	66
Arizona	21	61	71	63
Arkansas	14	13	12	16
California	0	0	1	3
Florida	43	60	50	56
Georgia	30	38	44	58
Louisiana	15	38	60	61
Mississippi	35	43	58	67
Missouri	0	1	1	2
New Mexico	1	6	35	23
North Carolina	6	3	12	20
Oklahoma	6	8	8	21
South Carolina	14	31	39	81
Tennessee	1	3	19	68
Texas	2	5	5	8
Virginia	0	0	3	7
US	13	17	21	29

The adoption of *Bt* varieties was extremely rapid in some areas and has been slower in others. After a year of very high budworm populations and damage in 1995, growers in Alabama adopted the new technology at an extremely rapid rate, planting over 60% of total acreage to *Bt* varieties in 1996. In 2000, 65% of cotton acreage in Alabama was in *Bt* varieties (USDA AMS).

Florida and Mississippi also adopted the technology on over 30% of cotton acreage in 1996. By 2000, four states had adoption rates of over 75%: Florida, Louisiana, Mississippi and Tennessee (USDA AMS).

Two major cotton producing states have had very low adoption rates thus far, which skews the national adoption rate. Texas, which accounted for 40% of cotton acreage in 2000, has only adopted *Bt* cotton on a small scale, at only 10% in 2000.

Adoption has been hindered by the lack of stripper varieties appropriate for growing conditions in Texas. California also has low adoption rates, which is at least partially due to a unique law in that state that controls which varieties may be planted. Demand for *Bt* cotton varieties in California is expected to be relatively low, however, since most producing areas of California are not infested with the three target pests of *Bt* varieties.

The impact of the adoption of Bt cotton varieties has been a reduction in yield losses due to Bt-target pests, reductions in insecticide use, and cost savings. For Alabama growers, yield losses due to bollworm/budworm were drastically reduced from levels experienced in 1995. In Central Alabama, where the average yield loss due to bollworm/budworm was 55% in 1995, losses varied between 2 and 7% from 1996 to 1999. The number of insecticide applications per year was reduced from 10 to between 0 and 2 (Williams). Yield losses due to Bt target pests have declined in 12 of 16 reporting states. The yield losses due to cotton bollworm, tobacco budworm and pink bollworm by state for 1985 to 1999.

The major impact of the adoption of *Bt* varieties is a reduction in insecticide use. There are two sources of cotton insecticide use data available. The first is the Cotton Council survey, which estimates insecticide use by target pest. The number of insecticide treatments for cotton bollworm, tobacco budworm and pink bollworm decreased in 1999 compared to 1995 in 12 of 16 reporting states. The number of insecticide applications directed towards cotton bollworm, tobacco budworm and pink bollworm by state from 1986 to 1999.

As part of the Cotton Council survey, three states report insecticide use separately for Bt acreage and conventional acreage. In all three states, the number of insecticide treatments directed at Bt target pests was lower on Bt acreage. However, treatments for other insects (except for boll weevil) were higher on Bt acreage.

This is believed to be due to the emerging importance of pests that were previously considered secondary pests, which were previously controlled with treatments at Bt target pests. However, the reduction in treatments for Bt targets is larger than the increased number of treatments towards these secondary pests in all three instances.

Finally, in Louisiana and Tennessee, the number of treatments for boll weevil is higher on Bt acreage than on conventional acreage. This is likely due to higher adoption rates for Bt cotton in boll weevil eradication areas. USDA pesticide use data record insecticide use by active ingredient without indication of target pest.

Comparing USDA pesticide use data for 6 states for 1995 to 1998 and 1999 shows a dramatic reduction in the use of insecticides used to control cotton bollworm, tobacco budworm and pink bollworm. The reductions in insecticide use for those insecticides that are recommended for cotton bollworm, tobacco budworm and pink bollworm control.

This difference in use was adjusted to account for reduction in planted acreage since 1995. The use of cotton insecticides for Bt target pests declined by over 2 million pounds between 1995, the year before Bt varieties were introduced, and 1998, and by 2.7 million pounds by 1999.

These reductions account for 10% and 14% of all insecticides used in those six states in 1995. The number of insecticide applications has also declined, by 8.7 million applications in 1998 and by 15 million applications in 1999, or 13% and 22% of the total number of insecticide applications in those states in 1995.

Several surveys have found that growers are achieving higher yields and attaining higher profits by planting *Bt* varieties, due to better pest control and decreased insect control costs.

The average increase in net returns from 5 studies in 7 states including 91 comparisons in 1999, comparing *Bt* to conventional varieties was $20.81/acre, taking into account the technology fee. On average, per acre insect control costs were $14.28 higher. This increased cost was outweighed by a yield increase of 9%, or 55 lbs/acre.

These yield and revenue impacts, if realized over all 4.67 million acres of *Bt* cotton in 1999, would result in a $99 million increase in revenues and 260 million lbs increased cotton production.

Herbicide Resistant Cotton

In 1995, BXN cotton was introduced. BXN cotton is tolerant to bromoxynil (Buctril), a post-emergence broadleaf herbicide already registered for use in corn and small grains. Roundup Ready cotton varieties became available in 1997, presenting growers with another option for post-emergence weed control using glyphosate (Roundup).

Nearly all U.S. cotton acreage is currently treated with herbicides. In 1998, 95% of U.S. cotton acreage was treated with herbicides (USDA NASS). The average number of herbicide applications per treated acre was 2.6 in 1995, while 34% of the acreage received 3 or more applications.

An average of 2.7 different active ingredients per acre were used in 1995, with 24% of the treated acreage receiving 4 or more active ingredients (USDA ERS 1997). Primary weed species are morningglory (*Ipomoea* spp.), pigweed (*Amaranthus* spp.), nutsedge (*Cyperus* spp.), cocklebur (*Xanthium* spp.) and johnsongrass (*Sorghum* spp.) (Byrd). Prior to 1995, cotton growers did not have any broadleaf herbicides that could be used over the top of a growing cotton crop without the potential to cause crop injury.

Instead, growers would use directed, post-emergence applications of nonselective herbicides, using specialized equipment to avoid herbicide contact with the growing crop plants, and cultivation. Directed, post-emergence treatments require weeds to be shorter than the cotton crop.

This height differential is sometimes difficult to achieve. Post-directed treatments on small cotton require time-consuming treatments and can damage plants if herbicides contact the plant foliage.

6

Elements of Commercial Agriculture

Although elements of the above ideal type of a competitive food system could be found in parts of the United States at various times, it never did exist in many areas of the country. For an individual farmer, the question was not how many firms were involved in the different stages of the food chain across the country, but rather how much of this commodity chain was accessible to his (sometimes her) farm.

AS farmers moved west, one of their major problems was how to transport their products to markets in the eastern cities. The government, wishing to promote increased industrialization, also perceived the problem and subsidized the construction of transportation systems, especially railroads. This often made the farmer dependent on a monopoly which could exploit him/her because of the unequal balance of economic power.

If a farmer had access to only one railroad, the power relationship certainly favoured the railroad. That farmer faced a monopoly regardless of how many other railroads existed in the country. Thus, many farmers faced the issue of monopoly control of capital from the time they became commercial farmers and began to be dependent on a single transportation system to move their products to the market.

Railroads in some parts of the country needed the business of farmers, but they had access to hundreds, if not thousands, of farmers. They were not dependent on any single farmer.

Their only concern was that farmers might be successful in organizing a united stand against the railroads. Frank Norris' agrarian populist novel, The Octopus: A History of California (1901), was about a handful of farmers who attempted such a united stand, rising up in direct rebellion, only to be dispatched by the railroads.

The whole history of the farmers' movement is largely about the unequal power balance between farmers and the railroads, and, more generally, between the farmers and all the "middlemen" they depended on for transportation, markets, and a host of inputs such as credit and farm equipment. In southwestern Minnesota, for example, the selling of water fowl to cities to the East and later the selling of agriculture products, especially grain, all depended on the railroad for transportation.

In the case of grain, the farmers also depended on the elevators (large silos) to store and to transfer it from farmers' wagons to the train. Many of the transnational corporations (TNCs) of today, like Cargill, exercised economic power in many of the local markets in which they began operating. Most farmers' movements were not successful in establishing alternative economic systems or firms which benefited the farmers, but with the help of legislation to encourage farmer cooperatives, there were some successes.

AGRICULTURE IN INDIA: ISSUES AND CHALLENGES

"Agriculture is the backbone of the Indian Economy"- said Mahatma Gandhi five decades ago. Even today, as we enter the new millennium, the situation is still the same, with almost the entire economy being sustained by agriculture, which is the mainstay of the villages. Not only the economy, but also every one of us looks up to agriculture for our sustenance too.

SIGNIFICANCE OF AGRICULTURE

Although agriculture contributes only 21% of India's GDP, its importance in the country's economic, social, and political fabric goes well beyond this indicator. The rural areas are still home to some 72 percent of the India's 1.1 billion people, a large number of whom are poor. Most of the rural poor depend on rain-fed agriculture and fragile forests for their livelihoods.

The sharp rise in foodgrain production during India's Green Revolution of the 1970s enabled the country to achieve self-sufficiency in foodgrains and stave off the threat of famine. Agricultural intensification in the 1970s to 1980s saw an increased demand for rural labour that raised rural wages and, together with declining food prices, reduced rural poverty.

Sustained, although much slower, agricultural growth in the 1990s reduced rural poverty to 26.3 percent by 1999/00. Since then, however, the slowdown in agricultural growth has become a major cause for concern. India's rice yields are one-third of China's and about half of those in Vietnam and Indonesia. With the exception of sugarcane, potato and tea, the same is true for most other agricultural commodities.

The Government of India places high priority on reducing poverty by raising agricultural productivity. However, bold action from policymakers will be required to shift away from the existing subsidy-based regime that is no longer sustainable, to build a solid foundation for a highly productive, internationally competitive, and diversified agricultural sector.

Issues and Challenges

It is here the challenge arises considering the implementation of the technology at various levels in the Global community. The need of the hour is not application of the technology but the adoption of appropriate technology, which would suit the particular level of the global community. In India, the farming practices are too haphazard and non-scientific and hence need some forethought before implementing any new technology.

Applications of agricultural inputs at uniform rates across the field without due regard to in-field variations in soil fertility and crop conditions does not yield desirable results in terms of crop yield.

The management of in-field variability in soil fertility and crop conditions for improving the crop production and minimizing the environmental impact is the crux of precision farming. Thus, the information on spatial variability in soil fertility status and crop conditions is a pre-requisite for adoption of precision farming.

Space technology including global positioning system (GPS) and GIS holds good promise in deriving information on soil attributes and crop yield, and allows monitoring seasonally- variable soil and crop characteristics, namely soil moisture, crop-phenology, growth, evapotranspiration, nutrient deficiency, crop disease, and weed and insect infestation, which, in turn, help in optimizing inputs and maximizing crop yield and income.

Though widely adopted in developed countries, the adoption of precision farming in India is yet to take a firm ground primarily due to its unique pattern of land holdings, poor infrastructure, lack of farmers' inclination to take risk, socio-economic and demographic conditions.

Factors Contribution to Decline of Agriculture

Slow Down in Agricultural and Rural Non-Farm Growth: Both the poorest as well as the more prosperous 'Green Revolution' states of Punjab, Haryana, Andhra Pradesh and Uttar Pradesh have recently witnessed a slow-down in agricultural growth and it ultimately lead for farmer's suicide. Some of the factors hampering the revival of growth are:

- Poor composition of public expenditures: Public spending on agricultural subsidies is crowding out productivity-enhancing investments such as agricultural research and extension, as well as investments in rural infrastructure, and the health and education of the rural people. In 1999/2000, agricultural subsidies amounted to 3 percent of GDP and were over 7 times the public investments in the sector.
- Over-regulation of domestic agricultural trade: While economic and trade reforms in the 1990s helped to improve the incentive framework, over-regulation of domestic trade has increased costs, price risks and uncertainty, undermining the sector's competitiveness.

- Government interventions in labour, land, and credit markets: More rapid growth of the rural non-farm sector is constrained by government interventions in factor markets — labour, land, and credit — and in output markets, such as the small-scale reservation of enterprises.
- Inadequate infrastructure and services in rural areas. Infrastructure is also a significant factor in the process of development but country like our rural Bharat has not posses the infrastructure such as roads, electricity, fertilizer and pesticides availability which caused the vulnerable damage to the growth of agriculture.
- Weak Framework for Sustainable Water Management and Irrigation: Inequitable allocation of water: Many states lack the incentives, policy, regulatory, and institutional framework for the efficient, sustainable, and equitable allocation of water.
- Deteriorating irrigation infrastructure: Public spending in irrigation is spread over many uncompleted projects. In addition, existing infrastructure has rapidly deteriorated as operations and maintenance is given lower priority.
- Inadequate Access to Land and Finance: Stringent land regulations discourage rural investments: While land distribution has become less skewed, land policy and regulations to increase security of tenure (including restrictions or bans on renting land or converting it to other uses) have had the unintended effect of reducing access by the landless and discouraging rural investments.
- Computerization of land records has brought to light institutional weaknesses: State government initiatives to computerize land records have reduced transaction costs and increased transparency, but also brought to light institutional weaknesses.
- Rural poor have little access to credit: While India has a wide network of rural finance institutions, many of the rural poor remain excluded, due to inefficiencies in the formal finance institutions, the weak regulatory framework, high transaction costs, and risks associated with lending to agriculture.
- Weak Natural Resources Management: One quarter of India's population depends on forests for at least part of their livelihoods.
- A purely conservation approach to forests is ineffective: Experience in India shows that a purely conservation approach to natural resources management does not work effectively and does little to reduce poverty.
- Weak resource rights for forest communities: The forest sector is also faced with weak resource rights and economic incentives for communities, an inefficient legal framework and participatory management, and poor access to markets.

- Weak delivery of basic services in rural areas: Low bureaucratic accountability and inefficient use of public funds: Despite large expenditures in rural development, a highly centralized bureaucracy with low accountability and inefficient use of public funds limit their impact on poverty. In 1992, India amended its Constitution to create three tiers of democratically elected rural local governments bringing governance down to the villages. However, the transfer of authority, funds, and functionaries to these local bodies is progressing slowly, in part due to political vested interests. The poor are not empowered to contribute to shaping public programmes or to hold local governments accountable.

Measures Needed Areas

- Enhancing agricultural productivity, competitiveness, and rural growth Enhancing productivity:
  - Creating a more productive, internationally competitive and diversified agricultural sector would require a shift in public expenditures away from subsidies towards productivity enhancing investments.
  - Second it will require removing the restrictions on domestic private trade to improve the investment climate and meet expanding market opportunities.
  - Third, the agricultural research and extension systems need to be strengthened to improve access to productivity enhancing technologies. The diverse conditions across India suggest the importance of regionally differentiated strategies, with a strong focus on the lagging states.
- *Improving Water Resource and Irrigation/Drainage Management:* Increase in multi-sectoral competition for water highlights the need to formulate water policies and unbundle water resources management from irrigation service delivery.

  Other key priorities include: (i) modernizing Irrigation and Drainage Departments to integrate the participation of farmers and other agencies in irrigation management; (ii) improving cost recovery; (iii) rationalizing public expenditures, withpriority to completing schemes with the highest returns; and (iv) allocating sufficient resources for operations and maintenance for the sustainability of investments.
- *Strengthening rural non-farm sector growth:* Rising incomes are fueling demand for higher-value fresh and processed agricultural products in domestic markets and globally, which open new opportunities for agricultural diversification to higher value products (*e.g.* horticulture, livestock), agro-processing and related services. The government

needs to shift its role from direct intervention and overregulation to creating the enabling environment for private sector participation and competition for agribusiness and more broadly, the rural non-farm sector growth. Improving the rural investment climate includes removing trade controls, rationalizing labour regulations and the tax regime (*i.e.* adoption of the value added tax system), and improving access to credit and key infrastructure (*e.g.* roads, electricity, ports, markets).

- Improving access to assets and sustainable natural resource use:
  - *Balancing poverty reduction and conservation priorities:* Finding win-win combinations for conservation and poverty reduction will be critical to sustainable natural resource management. This will involve addressing legal, policy and institutional constraints to devolving resource rights, and transferring responsibilities to local communities.
  - *Improving access to land:* States can build on the growing consensus to reform land policy, particularly land tenancy policy and land administration system. States that do not have tenancy restrictions can provide useful lessons in this regard. Over the longer term, a more holistic approach to land administration policies, regulations and institutions is necessary to ensure tenure security, reduce costs, and ensure fairness and sustainability of the system.
  - *Improving access to rural finance:* It would require improving the performance of regional rural banks and rural credit cooperatives by enhancing regulatory oversight, removing government control and ownership, and strengthening the legal framework for loan recovery and the use of land as collateral. It would also involve creating an enabling environment for the development of micro-finance institutions in rural areas.
  - Strengthening institutions for the poor and promoting rural livelihood.
  - *Promoting Community-Based Rural Development*: State Government efforts in scaling up livelihood and community-driven development approaches will be critical to build social capital in the poorest areas as well as to expand savings mobilization, promote productive investments, income generating opportunities and sustainable natural resource management. Direct support to self-help groups, village committees, user's associations, savings and loans groups and others can provide the initial 'push' to move organizations to higher level and access to new economic opportunities. Moreover, social mobilization and particularly the

empowerment of women's groups, through increased capacity for collective action will provide communities with greater "voice" and bargaining power in dealing with the private sector, markets and financial services.

- *Strengthening Accountability for Service Delivery:* As decentralization efforts are pursued and local governments are given more prominence in the basic service delivery, the establishment of accountability mechanisms becomes critical.

  Local governments' capacity to identify local priorities through participatory budgeting and planning needs to be strengthened. This, in turn, would improve the rural investment climate, facilitating the involvement of the private sector, creating employment opportunities and linkages between farm and non-form sectors.

FARM LEVEL DIVERSIFICATION IN DRYLAND REGIONS

Plot diversification arises from farmers owning more than one field, leasing in land, sub-ploting within fields, and multiple cropping over cropping season. More land fragmentation implies greater plot diversification.

Since, majority of farmers in dryland regions are small and marginal farmers, they have to meet multiple objectives with limited land.

Hence, they plan their cropping activities in such a manner that it can ensure them regular income, though low, and minimise risk of crop failure. Land fragmentation has many inherent limitations as there is a feeling that it leads to slow adoption of new technology.

But for farmers especially marginal and small, it is risk management strategy for spreading risk to larger areas.

Some a studies have been conducted based on country or district level data using gross revenue variability as a measure of crop income instability but they often fail to capture real farm level realities.

For example, the study by Barah and Binswanger (1982) using district level data for four states indicates that crop diversification was largely a response to low and unstable rainfall.

Moreover crop diversification does significantly reduce gross revenue variability. The effect of diversification has not been adequately addressed with farm level data on crop income stability.

The present study uses farm level data from three distinct agroclimatic regions covering a period of five years to address the causes and consequences of inter-regional and inter-village differences in diversification behaviour.

This data set from the ICRISAT Village Studies pertain to a panel of 30 cultivator households in each of the six villages located in three broad soil, climatic and cropping regions of the semi-arid tropics of India.

DETERMINANTS OF CROP DIVERSIFICATION

The level of crop diversification, both within and across villages, largely depends on site specific ecological characteristics, the resource endowment of farm households and their personal characteristics, emerging weather conditions, crop rotational requirements, perceived consumption needs of the households and market access.

However, all these factors such as crop rotational thoughts and consumption targets were not readily available with the data and so they have not been analysed.

It was expected that personal characteristics, including risk aversion lead to greater diversification. Risk attitudes for each household were estimated by experimental games described in Binswanger (1978) and Binswanger *et al* (1982).

Changing weather conditions particularly at planting is expected to influence crop choices and levels of diversification. The effect should be stronger in less assured rainfall regions where crops are planted in post-rainy season on residual soil moisture.

The resource endowments of the farm household is expected to have a substantial influence on diversification. The access to irrigation, availability of draft power, family labour, land quality, experience and education of the farmers, etc, play an important role in influencing diversification.

Since these three regions have diverse agroclimatic conditions the determinants of farm-level diversification are analysed separately for each region. A regional analysis is supported by F test that rejects at the 5% level, the hypothesis that the estimated regressions coefficients are equal across the three regions.

The regression results indicate that draft power availability is an important explanatory of variation in crop diversification across households within some of the regions. As expected, larger farms with more gross cropped area are more diversified than their smaller counterparts.

In contrast, inter household differences in family size have little bearing on variation in the levels of diversification. In Sholpur region, resource endowment thoughts are less important in explaining diversification which is largely accounted for by differences in land quality, cropping-year conditions, and village disparities within the region.

The influence of irrigated area is also significant and site specific. Limited well irrigation, particularly in the post-rainy season opens up opportunities to grow more crops such as wheat, chickpea, and other pulses.

Differences in risk aversion are strongly and positively associated with crop diversification in one region but in other regions do not play a strong role in determining crop diversification. Cropping year and site specific village characteristic influences have a negligible impact on crop diversification in the same region. However, in drought prone area of Sholapur crop diversification appears to be sensitive to changing weather conditions.

CHANGE AND HOUSEHOLD NET CROP INCOME

It is almost taken for granted in the risk literature that increasing diversification beyond a certain level leads to reduced risk at the cost of expected net crop income.

The size of such a trade-off is an empirical question. Household net crop income is defined as returns to land, capital and management, *i.e.* the cost of all purchased inputs and the value of family labour and owned draft power are deducted from the value of production. Household net crop income averaged ₹ 2,150, 2,190, 2,370 over the five cropping years for Mahbubnagar, Sholapur, and Akola regions respectively.

Household net cropping-year income is regressed on crop and plot diversification, land quality (in the Sholapur region), irrigated area, land value, gross cropped area, age and education.

Cropping-year dummies and within-region village variables are included to capture the average effects of changes in net crop income over time and between villages in the same region.

The diversification indices, GDI and PDI, are also squared and included in the regression to capture possible non-linear effects of diversification on income. Because the effect of the resource base on farm level diversification and ultimately on crop income is sequential, the inclusion of resource base determinants and levels of diversification as independent variables should not lead to simultaneity bias with ordinary least squares estimation. Nonetheless, it is better to draw inferences of association rather than causality from the estimated coefficients.

The importance of land quantity and quality in determining net crop income. In the Mahbubnagar villages, the significant quadratic response between crop diversification and income suggests that substitution away from the dominant crops particularly paddy does result in lost income. In Sholapur and Akola crop diversification does not appear to have reached a level that results in reduced income.

With the possible exception of Akola, the results suggest that fragmentation has not yet reached a level where inefficiencies associated with scattered holdings and small plot size noticeably detract from crop income.

The significant non-linear effects of plot diversification on income in Mahbubnagar are puzzling but may be explained by some complicated interactions between multiple cropping which is positively associated with plot diversification and irrigation. These effects may not have been fully captured by the irrigated area variable.

DIVERSIFICATION AND CROP INCOME STABILITY

The effectiveness of diversification as a self-insurance strategy. Stability is measured in a relative context and is identified with a low coefficient of variation for household crop income.

Crop income is measured in two ways, gross returns and net returns to family owned resources which include land, capital, family labour and management. All non-marketed outputs and inputs are valued at their respective opportunity costs in measuring gross and net returns.

Of the 180 cultivator households in the VLS sample, 134 have data on crop income for each of the five years from 1975-76 to 1979-80. The majority of the other 46 households leased out their land for one or more cropping years; a minority left the sample and were replaced.

Therefore, there are 134 cultivators for whom coefficients of variation of household gross and net crop returns are estimated from 1975-76 to 1979-80. CVs are calculated for each household across the five cropping years. It was also expected that crop income stability is a function of some of the resource base determinants.

Moreover, differences in crop income stability among farmers can also be attributed to fluctuations in the resource base across cropping years. To assess the impact of diversification on crop income stability, it is important to control for shifts in cultivated area, irrigated area, and land quality over time. This is particularly true for regions such as Sholapur where tenancy is prevalent (Jodha, 1981).

Descriptive information on the CVs' of the dependent and independent variables. The size of the CVs' at about 40 to 50% is another indication of the riskiness of dryland agriculture in the semi-arid tropics of peninsular India. Fluctuations in gross returns are somewhat less than in net returns. Somewhat surprisingly, both average gross and net returns are less stable in the more heavily irrigated Mahbubnagar region compared to the dryland Sholapur and Akola regions.

It is believed that crop income stability can be traced not only to the level of the resource base but also to its intertemporal variation. Because crop and plot diversification are themselves functions of the same resource base viewed in the context of association rather than causation.

It is perhaps heroic to try to explain inter-farm variation in crop income stability within a narrow ecological region based on only five year data. Nonetheless, some of the hypothesized determinants behave as expected and have statistically significant estimated coefficients.

Crop diversification is negatively and strongly associated with crop income instability in the more dryland Sholapur and Akola regions. A 10 percent proportional increase in the crop diversification index is associated with 4.8

and 13.0 decrease in the CV of net crop returns in Sholapur and Akola respectively. At the margin, crop diversification is about three times more effective in stabilizing net returns in rainfall assured Akola than in drought-prone Sholapur.

In Mahbubnagar, plot diversification contributes positively to crop income stability. In contrast, in the Akola and Sholapur villages, the results mildly suggest that plot diversification is associated with greater crop income instability. Production conditions for a crop in these more dryland villages must be quite narrowly defined as diversification across space with the same commodity does not dampen fluctuations in income.

This is particularly true in Sholapur where field diversification is positively correlated with gross and net return instability.

These results support Walker and Singh who found land sharecropped in Sholapur was of inferior quality and was more likely to be the site of crop failure than owner-operated land. Plot diversification appears to be a resource management strategy rather than an action that offers scope for crop income protection.

As expected, the CVs' of the resource base determinants are for the most part-the exception being land value in Mahbubnagar-positively associated with crop income instability. Larger farms having more gross cropped area are characterized by greater crop income stability in the three regions. However, the impact of farm size on crop income stability is not large.

Evaluated at their arithmetic means, a 10 percent increase in gross cropped area leads on are average to about a 2 percent reduction to the CV of net returns.

The evidence with respect to irrigation is not as clear, but it weakly suggests that greater access to irrigation is accompanied by greater instability in net returns particularly in the Akola villages where irrigated land accounts for only 3 percent of the gross cropped area. Besides following inter-regional differences were also observed:

- The impact of irrigation on crop diversification and income stability hinges upon cropping conditions and the existing level of irrigation. When an expansion in irrigation takes place for rainy season paddy production in a region where many farmers have some access to irrigation, irrigation leads to more specialization in paddy production. In contrast, when an investment in well irrigation occurs in a multicrop predominantly dryland region, it enhances opportunities for further crop diversification particularly in the postrainy season. More profitable production opportunities in turn translate into greater relative variance in net crop returns income and increased crop income instability. The net effect of irrigation on crop diversification and income stability depends on the source of irrigation, ecological location, and level of growth of irrigation facilities.

- Crop and plot diversification fluctuate more in more drought prone regions like Sholapur. Crop diversification appears to be more sensitive to agro-climatic determinants than plot diversification.
- Different regional ecological conditions offer differential opportunities for risk averse farmers to diversify. Therefore, it is not surprising to find that risk attitudinal differences vary regionally in their effect on diversification. Where they are significant, differences in risk attitudes do not exert a sizeable influence on decisions on diversification; their proportional impact is about the same as differences in draft power availability.
- Crop diversification effectively imparts stability to net crop returns in the dryland farming Sholapur and Akola villages. In the more highly irrigated Mahbubnagar villages, income instability and crop diversification are not significantly associated.

The implications of these conclusions are threefold.

First, at present levels of technology it appears that increasing land fragmentation associated with plot diversification does not represent a significant cost in terms of income foregone or a material benefit in the form of enhanced crop income stability.

Thus a prospective land consolidation programme would not compromise the risk-bearing capacity of farmers, nor would it significantly increase average crop income of farm households at existing levels of technology.

These issues require closer scrutiny within a more rigorous household economics framework.

Secondly, the emergence of crop diversification as a strong force in protecting net crop returns in the more rainfall-assured Akola region decreases the demand for government risk-reducing policies, such as crop insurance, because farmers can rely on self-insurance through diversification to enhance crop income stability.

These results further underscore the emerging consensus that the potential benefits of crop insurance programmes will be greater in more risk-prone, dryland regions like Sholapur for three reasons.

First, yield variability is the predominant source of risk in these regions (Barah and Binswanger, 1982), yields are more likely to be positively covariate across farmers which augments the efficiency of a homogeneous area approach to dampen fluctuations in crop income and third, there is less scope for crop diversification to be an effective self-insurance measure.

Lastly, an investment in irrigation may have a negative effect on relative crop income stability when instability is measured with respect to the coefficient of variation.

The farm-level effects of irrigation may be markedly different from what obtains at the district, state or national levels.

GROWTH OF DAIRY FARMING

We now pass to another very important area of agricultural capitalism in Russia, namely, the region in which not cereal, but livestock produce is of predominant significance. This region embraces, apart from the Baltic and the western gubernias, the northern, the industrial and parts of some of the central gubernias (Ryazan, Orel, Tula, and Nizhni Novgorod). Here animals are kept for dairy produce, and the whole character of agriculture is adapted to obtaining as large a quantity as possible of the more valuable market produce of this sort.

"Before our very eyes a marked transition is taking place from stock farming for manure to stock farming for dairy produce; it has been particularly notice able during the past ten years" (work quoted in previous footnote, *ibid.*). It is very difficult by the use of statistics to describe the various regions of Russia in this respect, because it is not the total number of horned cattle that is here important, but the number of dairy cattle and their quality. If we take the total number of animals per hundred inhabitants, we shall find that it is biggest in the outer steppe regions of Russia and smallest in the non-black-earth belt; we shall find that as time goes on the number diminishes.

Hence, we observe here what Roscher noted in his day, namely, that the number of animals per unit of the population is largest in districts of "extensive livestock farming". We, however, are interested in intensive livestock farming, and in dairy farming in particular.

We are compelled, therefore, to confine ourselves to the *approximate* computation made by the authors of the above-mentioned, *Sketch*, without claiming to make an exact estimate of the phenomenon; such a computation clearly illustrates the relative positions of the various regions of Russia as to degree of dairy-farm development. We quote this computation in *extenso*, supplementing it with some averages arrived at and data on the cheese-making industry in 1890 according to "factory" statistics.

To judge the development of dairy farming, we can only make use of data on butter production and cheese making. This industry arose in Russia at the very end of the 18th century (1795); cheese making on landlords' estates began to develop in the 19th century, but suffered a severe crisis in the 1860s, which opened the period of cheese making by peasants and merchants.

Thus, in 25 years production increased more than ten-fold; only the dynamics of the phenomenon may be judged from these data, which are extremely incomplete. Let us quote some more detailed material. In Vologda Gubernia an improvement in dairy farming began, properly speaking, in 1872, when the Yaroslavl-Vologda railway was opened; since then "farmers have begun to see to the improvement of their herds, to introduce grass cultivation, to acquire improved implements... and have tried to place dairy farming on a purely commercial basis".

In Yaroslavl Gubernia "the ground was prepared" by the so-called "cheese-making artels" of the 70s, and "cheese making continues to develop on the basis of private enterprise, merely retaining the title of 'artel'" ; cheese making "artels" figure—may we add—in the *Directory of Factories and Works* as establishments employing wage-workers. Instead of 295,000 rubles, the authors of the *Sketch* estimate the output of cheese and butter, according to *official* returns, at 412,000 rubles; correction of the figure brings the value of the output of fresh butter and cheese to 1,600,000 rubles, and if we add clarified butter and soft cheese, to 4,701,400 rubles, not counting either the Baltic or the western gubernias.

For the later period let us quote the following opinions from the above-cited publication of the Department of Agriculture *Hired Labour, etc*. Concerning the industrial gubernias in general we read: "A complete revolution in the position of the farms in this area has been brought about by the development of dairy farming"; it "indirectly has also helped to bring about an improvement in agriculture"; "dairy farming in the area is developing with every year".

In Tver Gubernia "there is to be observed the tendency both among private landowners and peasants to improve the methods of maintaining cattle"; the income from stock farming is estimated at 10 million rubles. In Yaroslavl Gubernia "dairy farming... is developing with every year.... Cheese and butter making have even begun to assume something of an industrial character... milk... is bought up from neighbours and even from peasants.

One comes across cheese factories run by a whole company of owners". "The general trend of private-landowner farming here," writes a correspondent from Danilov Uyezd, Yaroslavl Gubernia, "is marked at the present time by the following: 1) the transition from three-field to five-and seven-field crop rotation, with the sowing of herbage in the fields; 2) the ploughing up of disused lands; 3) the introduction of dairy farming, and as a consequence, the stricter selection of cattle and an improvement in their maintenance".

The same thing is said of Smolensk Gubernia, where the value of the output of cheese and butter amounted to 240,000 rubles in 1889—according to a report of the Governor (according to statistical returns, 136,000 rubles in 1890). The development of dairy farming is noted in the Kaluga, Kovno, Nizhni-Novgorod, Pskov, Esthland and Vologda gubernias.

The value of the output of butter and cheese in the last-mentioned gubernia was estimated at 35,000 rubles according to statistics for 1890, to 108,000 rubles according to the Governor's report, *and to 500,000 rubles* according to local returns for 1894, which gave a total of *389 factories*. "That is what the statistics say. Actually, however, there are far more factories, since, according to investigations by the Vologda Zemstvo Board, there are 224 factories in Vologda Uyezd alone." Production is developed in three uyezds, and has partly penetrated a fourth. One can judge from this how many times the above quoted figures need to be multiplied in order to approach the real situation. The plain view of an expert that at the present time the number of butter and cheese-making establishments "amounts to several thousand", gives a truer picture of the facts than the allegedly exact figure of 265.

Thus the data leave not the slightest doubt about the enormous development of this special type of commercial farming. The growth of capitalism was accompanied here too by the transformation of routine technique. "In the sphere of cheese making," we read, for example, in *Agriculture and Forestry*, "more has been done in Russia during the last 25 years than perhaps in any other country".

Mr. Blazhin says the same thing in his article "Technical Progress in Dairy Farming". The principal change is that the "age-old" method of leaving cream to settle has been replaced by the system of separating cream in centrifugal machines (separators).

The machine has enabled the work to be carried on irrespective of atmospheric temperature, increased the butter yield from milk by 10%, improved the quality of the product, reduced the cost of butter production (the machine requires less labour, space, and ice, as well as fewer utensils), and has led to the concentration of production.

Large peasant butteries have grown up, handling "as much as 500 poods of milk a day, which was physically impossible... when the milk was left to settle". Improvements are being made in the instruments of production (permanent boilers, screw presses, improved cellars), and production is being assisted by bacteriology, which is providing pure cultures of the type of lactic-acid bacilli needed for fermenting cream.

Thus, in the two areas of commercial farming we have described, the technical improvements called into being by the requirements of the market were effected primarily in those operations that were easiest to change and are particularly important for the market: reaping, threshing and winnowing in commercial grain farming, and the technical processing of animal produce in the area of commercial stock farming.

As to the keeping of cattle, capital finds it more profitable for the time being to leave that to the small producer: let him "diligently" and "industriously" tend "his" cattle, let him bear the brunt of the hardest and roughest work of tending the milk-yielding machine.

Capital possesses the latest improvements and methods not only of separating the cream from the milk, but also of separating the "cream" from this "diligence", of separating the milk from the children of the peasant poor.

COSMETIC THE PRODUCTIVITY OF SMALL FARMING SYSTEMS

Despite the evidence of the resiliency and productivity advantages of small-scale and traditional farming systems, many scientists and growth specialists and organizations argue that the performance of subsistence agriculture is unsatisfactory, and that agrochemical and transgenic intensification of production is essential for the transition from subsistence to commercial production. Although such intensification approaches have met with much failure, research indicates that traditional crop and animal combinations can often be adapted to increase productivity.

This is the case when ecological principles are used in the redesign of small farms, enhancing the habitat so that it promotes healthy plant growth, stresses pests, and encourages beneficial organisms while using labour and local resources more efficiently.

Several reviews have amply documented that small farmers can produce much of the needed food for rural and neighbouring urban communities in the midst of climate change and burgeoning energy costs.

The evidence is conclusive: new agroecological approaches and technologies spearheaded by farmers, NGOs, and some local governments around the world are already making a sufficient contribution to food security at the household, national, and regional levels. A variety of agroecological and participatory approaches in many countries show very positive outcomes even under adverse environmental conditions.

Potentials include: raising cereal yields from 50 to 200 percent, increasing stability of production through diversification, improving diets and income, and contributing to national food security (and even to exports) and conservation of the natural resource base and biodiversity.

This evidence has been reinforced by a recent report of the United Nations Conference on Trade and Growth stating that organic agriculture could boost African food security. Based on an analysis of 114 cases in Africa, the report revealed that a conversion of farms to organic or near-organic production methods increased agricultural productivity by 116 percent.

Moreover, a shift towards organic production systems has enduring impact, as it builds up levels of natural, human, social, financial, and physical capital in farming communities.

The International Assessment of Agricultural Knowledge, Science and Technology (AKST) commissioned by World Bank and the Food and Agriculture Organization (FAO) of the United Nations recommended that an increase and strengthening of AKST towards agroecological sciences will contribute to addressing environmental issues while maintaining and increasing productivity. The assessment also stresses that traditional and local knowledge systems enhance agricultural soil quality and biodiversity as well as nutrient, pest, and water management, and the capacity to respond to environmental stresses such as climate.

Whether the potential and spread of agroecological innovations is realized depends on several factors and major changes in policies, institutions, and research and growth approaches. Proposed agroecological strategies need to target the poor deliberately, and not only aim at increasing production and conserving natural resources. But they must also create employment and provide access to local inputs and local markets. Any serious attempt at developing sustainable agricultural technologies must bring to bear local knowledge and skills on the research process.

Particular emphasis must be given to involving farmers directly in the formulation of the research agenda and on their active participation in the process of technological innovation and dissemination through Campesino a Campesino models that focus on sharing experiences, intensification local research, and problem-solving capacities.

The agroecological process requires participation and enhancement of the farmer's ecological literacy about their farms and resources, laying the foundation for empowerment and continuous innovation by rural communities. Equitable market opportunities must also be developed, emphasizing local commercialization and distribution schemes, fair prices, and other mechanisms that link farmers more directly and with greater solidarity to the rest of the population.

The ultimate challenge is to increase investment and research in agroecology and scale up projects that have already proven successful to thousands farmers. This will generate a meaningful impact on the income, food security, and environmental well-being of all the population, especially small farmers who have been adversely impacted by conventional modern agricultural policy, technology, and the penetration of multinational agribusiness deep into the third world.

RURAL SOCIAL MOVEMENTS, AGROECOLOGY, AND FOOD SOVEREIGNTY

The growth of sustainable agriculture will require significant structural changes, in addition to technological innovation, farmer-to-farmer networks, and farmer-to-consumer solidarity.

The required change is impossible without social movements that create political will among decision-makers to dismantle and transform the institutions and regulations that presently hold back sustainable agricultural growth. A more radical transformation of agriculture is needed, one guided by the notion that ecological change in agriculture cannot be promoted without comparable changes in the social, political, cultural, and economic arenas that help determine agriculture.

The organized peasant and indigenous-based agrarian movements—such as the international peasant movement La Vía Campesina and Brazil's Landless Peasant Movement (MST)—have long argued that farmers need land to produce food for their own communities and for their country. For this reason they have advocated for genuine agrarian reforms to access and control land, water, and biodiversity that are of central importance for communities in order to meet growing food demands.

Vía Campesina believes that in order to protect livelihoods, jobs, people's food security, and health as well as the environment, food production has to remain in the hands of small-scale sustainable farmers and cannot be left under the control of large agribusiness companies or supermarket chains.

Only by changing the export-led, free-trade based, industrial agriculture model of large farms can the downward spiral of poverty, low wages, rural-urban migration, hunger, and environmental degradation be halted. Social rural movements embrace the concept of food sovereignty as an alternative to the neo-liberal approach that puts its faith in an inequitable international trade to solve the world's food problem. Instead, it focuses on local autonomy, local markets, local production-consumption cycles, energy and technological sovereignty, and farmer-to-farmer networks.

"Greening" the Green Revolution will not be sufficient to reduce hunger and poverty and conserve biodiversity. If the root causes of hunger, poverty, and inequity are not confronted head-on, tensions between socially equitable growth and ecologically sound conservation are bound to accentuate.

Organic farming systems that do not challenge the monoculture nature of plantations and rely on external inputs as well as foreign and expensive certification seals, or fair-trade systems destined only for agro-export, offer very little to small farmers that become dependent on external inputs and foreign and volatile markets.

By keeping farmers dependent on an input substitution approach to organic agriculture, fine-tuning of input use does little to move farmers towards the productive redesign of agricultural ecosystems that would move them away from dependence on external inputs. Niche markets for the rich in the North exhibit the same problems of any agro-export scheme that does not prioritize food sovereignty, perpetuating dependence and hunger. Rural social movements understand that dismantling the industrial agrifood complex and restoring local food systems must be accompanied by the construction of agroecological alternatives that suit the needs of small-scale producers and the low-income non-farming population, and that oppose corporate control over production and consumption.

Given the urgency of the problems affecting agriculture, coalitions that can rapidly foster sustainable agriculture among farmers, civil society organizations (including consumers), as well as relevant and committed research organizations are needed. Moving towards a more socially just, economically viable, and environmentally sound agriculture will be the result of the coordinated action of emerging social movements in the rural sector in alliance with civil society organizations that are committed to supporting the goals of these farmers movements. As a result of constant political pressure from organized farmers and others, politicians will, it is hoped, become more responsive to developing policies that will enhance food sovereignty, preserve the natural resource base, and ensure social equity and economic agricultural viability.

RAINFED FARMING DEVELOPMENT

The Rainfed agriculture plays an important role in Indian economy. In India 68 percent of total net sown area (136.8m.ha) comes under rainfed lands spread over 177 districts. Rainfed crops account for 48 percent area under food crops and 68 percent of the area under non-food crops.

Nearly 50 percent of the total rural workforce and 60 percent of livestock in the country are concentrated in the dry districts. In Tamil Nadu, out of 7 m.ha of cultivable area, around 3.1m.ha comes under rainfed agriculture.

But uncertainty in production due to fluctuations in total rainfall and changes in its distribution, decrease in relative productivity in rainfed lands etc. affect the livelihoods of many poor and marginalised farmers. Considering the importance and issues of rainfed farming, DHAN Foundation initiated an experimental project on Rainfed Agriculture with following objectives.

- Understanding the status and issues of rainfed farming in the locations selected and identification of specific interventions.
- Improving the total factor productivity of rainfed farming.

The new theme was launched on October 2, 2002 on the occasion of the DHAN Foundation Day celebrations. The pilot project is in the initial phase of implementation. In order to understand rainfed farming context in South India a farmers workshop was organised on the theme *'Rainfed Farming – Past, Present and Future'*. Farmers and experts in this field attended the workshop. The focus of the workshop was on understanding the changes in rainfed farming in the last twenty years and issues faced by these farmers. The prevailing best practices in rainfed farming were also shared. The deliberations in the workshop were on generating ways and means to address the issues that had been identified.

After thorough consultation and screening, it was decided to experiment on this theme in two different agro climatic zones, one in Kallikudi block of Madurai district, which represents black soil. The second one is Natrampalli block of Vellore district, which represents red soil and hard rock areas. In this block dependence on groundwater is present in part of it, while the rest is predominantly drylands. The following areas have been taken up for experimentation in the rainfed farming development theme.

- Understanding and evolving suitable community organisation model
- Transfer of farmers best practices for wider use.
- Identification and transfer of suitable technologies evolved by the research institutions.

- Finding suitable extension processes and arrangements for the same.
- Finding solutions to issues not addressed by best practices and modern technologies.

Focus Areas

- Creating a community organisation base for the theme in the new location by designing a suitable Community Organisation model for rejuvenating rain fed farming.
- Understanding the rain fed farming practices in the location.
- Trying out few experiments to know their suitability to the location.
- Placing a team & building their capacity on the theme.

Progress

The rainfed farming development theme has evolved a people organisation model for piloting. Uzhavar kuzhu (Farmers groups) have been formed with 15 to 20 farmers who have lands at close proximity as members. 126 Uzhavar kuzhus have been promoted with 2039 members with both men and women. Kalanjiams are formed with landless agriculture labourers and for tenant farmers. 16 Kalanjiams have been promoted with 249 members. The team has promoted twelve watershed associations.

Situation specific and need-based interventions were attempted on experimental basis. These attempts were centred on the local best practices and indigenous technology.

Focus was given on individual farm that too with small and marginal farm holdings; where the vulnerability is very high. Landless farmers are given central focus, providing them access to lands is one of the key purposes of the theme.

The team has evolved several areas for observation and experimentation. The resources for the work have been mobilised through the NWDPRA (National Watershed Development Project for Rainfed Areas) programme. Observation studies have been taken up with 123 farming families. Experimentation on technologies suggested by the agricultural universities and the indigenous ones were attempted to find the suitability for the selected locations.

The areas in which experiments were taken up are improved agronomic practices, plant protection measure, crop improvement and farm mechanization. The farmers were trained to observe the performance of each experiment. Farmers could see positive results in some of the experiments. They are interested to follow them in the coming seasons.

The team has built good relationships with District Nodal Agency (Department of Agriculture). The team could undertake entry point activities like deepening of drinking water wells and construction of hand operated bore wells to solve the drinking water problems in the villages in collaboration with them. Drought proofing works were undertaken with the grant support from NOVIB, the Netherlands in both the locations.

The teams gained a lot of learning which was very useful to define the intervention areas for the theme. An Advisory board has been constituted with eminent persons with experience in this field at time of initiation of the theme.

The advisory board met once and the theme related progress including the observation and the experimentation studies were shared with them to get their feedback and suggestions. Several capacity building events were organised for the team members to build their knowledge and skills on community mobilisation and technical aspects.

The team undertook and completed two research studies namely a) Factors responsible for cropping pattern change in Thirumangalam Taluk of Madurai district and b) An analysis of conversion of dryland into garden land in Nattarampalli block of Vellore district. These two studies were useful in understanding the context of the locations.

CONVERTING LAND-USE AND FUTURE OF AGRICULTURE

One of the most important consequences of growing pressure on land is the declining trend in the average farm size and the pattern of holdings. According to the latest Agricultural Census in 1970-71 there were 70 million holdings operating 162 million ha. By 1990-91 there were 105 million holdings operating 165 million ha.

The average farm size decreased from 2.30 ha in 1970-71 to 1.57 ha in 1990-91. As of 1990-91 about 78 percent of holdings were small (1.0 to 2.0 ha) and marginal (<1.0 ha). A little more than 20 percent of the farmers were semi-medium (2.0 to 4.0 ha) and medium (4.0 to 10.0 ha).

Large farmers (>10.0 ha) constituted only 1.6 percent of the total holdings. Over the twenty-year period since 1970 the proportion of marginal farmer has increased from 50 to 59 percent and that of large farmer has declined from about 4 to 1.6 percent.

The proportion of total area operated by marginal farmers increased from nine percent in 1970-71 to nearly 15 percent in 1990-91 while the proportion of large farmers declined from about 31 percent to 17 percent in the same period. The size of average holding is very unevenly distributed among the states. States with relatively large average size of operational holding are a mixed lot – they include states with large tracts of barren lands *e.g.* Rajasthan, Maharashtra and Madhya Pradesh on the one hand and agriculturally advanced state like Punjab on the other.

These trends in farm size changes will have a profound effect on the future agricultural development strategies.

Although India's population growth rate has slowed from 2.1 percent in 1980s to 1.8 percent in the 1990s and is expected to slow further in the coming decades, yet the population is projected to reach 1.33 billion by 2020 from the current one billion.

The urban share of total population is projected to increase from 26 percent to 35 percent of the total population. Although incidence of poverty is falling, it is estimated that in 93-94 (upto which data is available) 320 million people constituting 36 percent of the population were below the officially defined poverty line.

The nature of the poverty line has been shifting. About 30 years ago 48.4 percent of those living in rural areas were poor and 20 percent of those living in the urban areas were classed as poor. Recent studies show that the number of poor in urban areas have been increasing at relatively higher rate compared to the rural areas.

At present those below the poverty line in rural sector constitute 37 percent of the population while in the urban sector the percentage is 32 percent. In the context of poverty alleviation, therefore, emphasis will be required to be placed both on production of food by the poor as well as on the availability of food for the urban poor.

It needs to be recognized that a large proportion of the rural poor are located in regions of low potential for food production *e.g.* arid and semi-arid areas, hilly regions, degraded land and forest areas. Widespread hunger and malnutrition are the direct manifestation of poverty and will call for increasing efforts to produce more food at affordable price.

Increasing population and economic growth are changing patterns of land use making potentially unsustainable demands on the country's natural resources.

- Since early fifties the net area sown was expanded rapidly at first but at a diminishing rate since 1970 to reach approximately 142 million ha at present. During 1950s and 1960s areas under agriculture expanded substantially as the fallows were reduced and cultivable wastes were put under the plough. The net area sown increased from 119 million ha in 1950-51 to 133 million ha by 1960-61 and further to 140 million ha by 1970-71. Fallow lands declined from 28 million ha in 1950-51 to 20 million ha by 1970-71. Cultivable wastelands declined from 23 to 17.5 million ha.
- Land use intensity *i.e.* fraction of net sown area to total geographical area increase from 36 percent in 1950-51 to 40.5 percent in 1960-61 and 43 percent by 1970-71 where it has since stabilized.
- Cropping intensity *i.e.* gross sown area as percent of net sown area increased from 111 percent in 1950-51 to 115 percent in 1960-61, 118 percent in 1970—71 and 130 percent by mid 1990s.

- While the contribution of increased area in the growth of agriculture has declined over time, that of productivity has increased. The yield of all crops grew at 1.5 percent per annum between early 1950s and mid 1960s. the pace accelerated to 1.7 percent in the 1970s and then to 3 percent per annum between early 1980s and mid 90s. Unlike the gains in area, which benefited non-foodgrains, the gains in productivity accrued mostly to foodgrains.
- India's forest resources have been dwindling. According to the 'State of Forest Report' (1997) the total forest cover of the country is estimated at 63.34 million ha *i.e.* 19.27 percent of the geographic area of the country. Of these the dense forest (crown density more than forty percent) and open forest (crown density 10 to 40 percent) occupying about 11 and 8 percent of the geographic area respectively and mangroves occupy 0.15 percent of the geographic area. The country has lost about 5482 sq. km. of forest cover since the 1995 assessment. By any estimate the area under forest is far below the national policy goals and many areas nominally under forest are being used for non-forest purposes. Similarly 'uncultivated lands' such as permanent pastures, miscellaneous tree crops, cultivable wastes and fallow is subject to increasing competition from uses other than feeding livestock.
- The growth of livestock population is an important source of competition for land.
- The area sown to fodder crops is not recorded. Information available from other sources provide an estimate ranging from 4 to 5.5 percent of the net sown area and suggest that the area under fodder crops will have to increase to 10 percent or more to support increasing livestock based activity.

The pressure on India's land and water resources is seriously threatening native plant and animal diversity. India has uniquely rich and diverse genetic base. With increasing agriculture and economic development the genetic pool is declining. This decline, if unchecked and poorly managed can have unforeseen and adverse consequences for the sustainability of agriculture of the region.

AGRICULTURE IN THE CHANGING WORLDWIDE SCENARIO

Steady globalization of trade has profound implications for future agricultural development. The diversity of India's agro-ecological setting, high bio-diversity and relatively low cost of labour provide potential for agricultural competitiveness in a globalized economy. It is expected that with increasing globalization of markets over the years there will be demands for agricultural intensification. This will also be favoured because of greater backward and forward linkages between agriculture and food industry. Therefore, increase in production and productivity are bound to be strategically important to economy. Intensification will not only favour alleviation of rural poverty but will also improve resource conservation particularly in the small farming sector where farmers can be encouraged to take up organized production of high value crops such as fruits, specialty vegetables, flowers medicinal and aromatic herbs etc. Stronger demands for crops of the small farmers' will not only improve incomes and welfare but will also make investments in technology and resource conservation more attractive.

The General Agreement on Tariff and Trade (GATT) and liberalization of global trade is bound to have impact on future land use and production pattern. Understanding the local, national and international environment under which agricultural production is taking shape will be crucial in developing our own strategies.

TECHNOLOGICAL NEEDS AND FUTURE AGRICULTURE

It is apparent that the tasks of meeting the consumption needs of the projected population are going to be more difficult given the higher productivity base than in 1960s. There is also a growing realization that previous strategies of generating and promoting technologies have contributed to serious and widespread problems of environmental and natural resource degradation.

This implies that in future the technologies that are developed and promoted must result not only in increased productivity level but also ensure that the quality of natural resource base is preserved and enhanced. In short, they lead to sustainable improvements in agricultural production.

Productivity gains during the 'Green Revolution' era were largely confined to relatively well endowed areas. Given the wide range of agroecological setting and producers, Indian agriculture is faced with a great diversity of needs, opportunities and prospects. Future growth needs to be more rapid, more widely distributed and better targeted. Responding to these challenges will call for more efficient and sustainable use of increasingly scarce land water and germplasm resources.

Technical solutions required to solve problems will be increasingly location-specific and matched to the huge agroecological/climatic diversity. Detailed indigenous knowledge and greater skills in blending modern and traditional technologies to enhance productive efficiency will be more than ever before, key to the farming success and sectoral growth. Most technological solutions will have to be generated and adapted locally to make them compatible with socio-economic conditions of farming community.

New technologies are needed to push the yield frontiers further, utilize inputs more efficiently and diversify to more sustainable and higher value cropping patterns. These are all knowledge intensive technologies that require both a strong research and extension system and skilled farmers but also a reinvigorated interface where the emphasis is on mutual exchange of information bringing advantages to all. At the same time potential of less favoured areas must be better exploited to meet the targets of growth and poverty alleviation.

These challenges have profound implications for products of agricultural research. The way they are transferred to the farmers and indeed the way research is organized and conducted. One thing is, however, clear – the new generation of technologies will have to be much more site specific, based on high quality science and a heightened opportunity for end user participation in the identification of targets. These must be not only aimed at increasing farmers' technical knowledge and understanding of science based agriculture but also taking advantage of opportunities for full integration with indigenous knowledge. It will also need to take on the challenges of incorporating the socio-economic context and role of markets.

With the passage of time and accelerated by macro-economic reforms undertaken in recent years, the Institutional arrangements as well as the mode of functions of bodies responsible for providing technical underpinning to agricultural growth are proving increasingly inadequate. Changes are needed urgently to respond to new demands for agricultural technologies from several directions. Increasing pressure to maintain and enhance the integrity of degrading natural resources, changes in demands and opportunities arising from economic liberalization, unprecedented opportunities arising from advances in biotechnology, information revolution and most importantly the need and urgency to reach the poor and disadvantaged who have been by passed by the green revolution technologies.

Another important implication of increasing globalization relates to the need for greater attention to the quality of produce and products both for the domestic and the foreign markets. This would imply that production must be tuned to actual rapidly changing product demand. Such adaptation to global markets would require state of the art research, which can be achieved only by setting global standards of research, focus on well defined priorities and mechanisms which permit close interaction of farmers with researchers, the private sector and markets.

Extension Strategies

Since early fifties a number of public by funded agricultural development programmes have been sponsored. These have included programmes like the National Extension Service (NES) Blocks in 1953, the Intensive Agricultural District Programme (IADP) in 1961-62, the Intensive Agricultural Area Programme (IAAP) 1964-65, the High Yielding Variety (HYV) programme 1966-

67 and the Small and Marginal Farmers' Development Programmes (SMFDP) in 1969-70. Though these programmes had a perceptible impact the efforts did not get replicated over different areas and categories of farmers. In mid seventies based on pilot level project in Rajasthan Canal and Chambal command area a 'Training and Visit' (T&V) system of extension was promoted in different states. Extension efforts of the Indian Council of Agricultural Research through its research Institutes and the State Agricultural University were largely limited to demonstration of new technologies through such programmes as National Demonstration Project, Operational Research Project, the Lab to Land Programme and the Krishi Vigyan Kendras. However, there appears much to be desired in the way that extension programmes are conceived and implemented.

At present extension programmes are implemented in largely a top-down fashion leaving little scope for localized planning and action. Farmers are almost passive receivers and their involvement in the process of technology generation and adoption is almost absent. Extension services, at present, are almost exclusively in the public sector domain and there is no effort or institutional support for other operators *e.g.* the NGOs, the corporate bodies etc.

Extension programmes sponsored by the government operate largely in isolation and there appears a strong need to view the extension programmes as an integral part of the research and development process.

7

Introduction to Horticulture and Plant Physiology

Horticulture is classically defined as the culture or growing of garden plants. Horticulturists work in plant propagation, crop production, plant breeding and genetic engineering, plant biochemistry, plant physiology, and the storage, processing, and transportation of fruits, berries, nuts, vegetables, flowers, trees, shrubs, and turf.

They improve crop yield, quality, nutritional value, and resistance to insects, diseases, and environmental stresses. Genetics is also used as a valuable tool in the development of plants that can synthesize chemicals for fighting disease (including cancers).

Horticulture involves five areas of study. These areas are floriculture (includes production and marketing of floral crops), landscape horticulture (includes production, marketing and maintenance of landscape plants), olericulture (includes production and marketing of vegetables), pomology (includes production and marketing of fruits), and postharvest physiology (involves maintaining quality and preventing spoilage of horticultural crops).

Horticulturists can work in industry, government, or educational institutions. They can be cropping systems engineers, wholesale or retail business managers, propagators and tissue culture specialists (fruits, vegetables, ornamentals, and turf), crop inspectors, crop production advisors, extension specialists, plant breeders, research scientists, and of course, teachers.

College courses that complement Horticulture are biology, botany, entomology, chemistry, mathematics, genetics, physiology, statistics, computer science, and communications, garden design, planting design. Plant science and horticulture courses include: plant materials, plant propagation, tissue culture, crop production, post-harvest handling, plant breeding, pollination management, crop nutrition, entomology, plant pathology, economics, and business.

Some careers in horticultural science require a masters (MS) or doctoral (PhD) degree. Horticulture takes place in many gardens and plant growth centres. Plants are often grown as seedlings within plant nurseries. Activities in nurseries range from preparing seeds and cuttings to growing fully mature plants. These are often sold or transferred to ornamental gardens or market gardens.

Horticulture is a broad field of study encompassing the production, maintenance, sales, and service of bedding, flowering, ornamental, and nursery plants and food crops.

This major has four distinct concentrations. Floriculture Crop Production stresses greenhouse production. General Horticulture allows students to design their own focus of study. Retail Floral Design emphasizes the principles and practices which prepare students for the successful operation of a retail floral business by concentrating on design, handling of plants, and business practices. Management of Woody Plants covers plant production and management in a nursery and the operation of a retail garden centre.

Increasingly precise management of horticultural crops lends itself to the use of mechanistic crop simulation models to predict growth and yield. In this chapter, we encourage systematic approaches for developing, parameterizing, and testing of dynamic crop growth models for horticultural crops.

Using the *Cropgro* model as an example, we describe crop development, photosynthesis, leaf area expansion, as well as the addition, growth, and maturation of individual fruits.

An algorithm is described for predicting fruit fresh weight, size, and dry matter concentration from dry matter growth and physiological age of individual tomato (Lycopersicon esculentum Mill.) fruits. Parameterizing crop models for response to climatic factors, especially temperature, requires interpretive use of literature information, along with testing against growth data from a range of growth conditions. A review of tomato literature for temperature dependencies of processes is presented with a view to how this information would be used to parameterize the CROPGRO-tomato model.

In modern horticulture, growers are searching for reliable biosensors and diagnostic tools to evaluate plant eco-physiological behaviour in order to optimize plant growth and crop development. One of these biosensors is the linear variable displacement transducer (LVDT) which accurately measures stem diameter variations.

The variations of stem diameter reflect the combination of two main phenomena: irreversible radial stem growth and daily shrinkage and swelling of the elastic stem cells due to changes in internally stored water. A correct interpretation of the composite LVDT-signal, with respect to its practical application, requires an unambiguous distinction between these two components of the signal.

This chapter describes a comprehensive model for stem diameter variations which includes a mechanistic description of radial stem growth and elastic stem diameter changes related to changes in internally stored water. Using the results of a two-year-old beech tree (Fagus sylvatica), the practical application of the model is illustrated with measured LVDT-signals where radial stem growth and elastic changes of the stem could be distinguished. As this approach is also applicable to other ornamental or vegetable crops, the model can be considered as a very powerful tool for interpretation of the composite and complex origin of the variations in stem diameter as is observed with LVDTs.

TROPICAL AND SUB-TROPICAL FRUITS

BANANA

Banana (Musa paradisiaca L.) occupies over 1,64,000 hectares, mainly in Tamil Nadu, West Bengal, Kerala, Maharashtra, Gujarat, Karnataka, Assam, Andhra Pradesh and Bihar. Though some inferior types of banana are found growing as far north as the Himalayas, its commercial importance is mainly limited to the more tropical conditions, such as those prevailing in central, southern and north-eastern India. It is a moisture- and heat-loving plant and cannot tolerate frost or arid conditions.

Varieties

Cultivated varieties are broadly divided into two groups: table and culinary. Among the former are 'Poovan' in Madras (also known as 'Karpura Chakkarekeli' in Andhra Pradesh); 'Mortaman', 'Champa' and 'Amrit Sagar' in West Bengal; 'Basrai', Safed Velchi', Lal Velchi' and 'Rajeli' in Maharashtra; 'Champa' and 'Mortaman' in Assam and Orissa; and 'Rastali', 'Sirumalai', 'Chakkarekeli', 'Ney Poovan', 'Kadali' and 'Pacha Nadan' in southern India. 'Basrai', which is known under different names, *viz.* 'Mauritius', 'Vamankeli', 'Cavendish', 'Governor', 'Harichal', is also grown in central and southern India. Recently, the 'Robusta' variety is gaining popularity in Tamil Nadu and Karnataka.

The 'Virupakshi' variety (Hill banana) is the most predominant variety in the Palni Hills of Tamil Nadu. Among the culinary varieties, Nendran bananas, 'Monthan', 'Myndoli' and 'Pacha Montha Bathis' are the leading commercial varieties in southern India. 'Gros Michel' is a recent introduction into southern India; it is suitable for cultivation only under garden-land conditions and is generally fastidious in its cultural requirements. It is not, therefore, in favour with the cultivation.

Propagation and Planting

Propogation is by suckers or off-shoots which spring at the base of a banana-tree from underground rhizomes. Vigorous suckers, with stout base, tapering towards the top and possessing narrow leaves, are selected for plant. Each sucker should have a piece of underground stem with a few roots attached to it.

Banana suckers can be planted throughout the year in southern India, except during summer, whereas in the rest of the country, the rainy season is preferred. They are planted in small pits, each just enough to accommodate the base of a sucker. The planting-distance varies from 2m×2m in the case of dwarf varieties to 4m×4m in the case of very tall varieties.

Manuring

An application of 20 to 25 kg of farmyard manure, together with about 5 kg of wood-ashes per plant is given at planting time. In southern India, ammonium sulphate is applied one month, five months and nine months after planting 20 kg per hectare each time.

In western India, a little over 2 kg of oilcake per stool is applied during the first three months after planting.

A complete fertilizer mixture may be applied to supply 100 to 200 kg of N, 100 to 200 kg of P_2O_5 and 200 to 400 kg of K_2O per hectare.

A green-manure crop is also considered beneficial. Trials at the Indian Institute of Horticultural Research have shown that for the 'Robusta' variety, a fertilizer mixture comprising 180 g of N + 108 g of P_2O_5 + 225 g of K_2O per plant is ideal.

After-Care

The removal of suckers, dry leaves and pseudostems, from which the fruits have been harvested, constitute the main after-care. Daughter-suckers should be removed promptly until the mother-plant flowers, when one daughter-sucker may be allowed to take its place. The removal of dry leaves and useless pseudostems requires to be done in time. After all the fruits are formed, the pendant portion of the remaining inflorescence along with the heart should be removed.

The propping of plants with bamboo poles, especially those which have thrown out bunches, is necessary wherever damage by wind is apprehended. Where the wind damage is recurring, dwarf varieties should be preferred.

Irrigation

The banana-plants require very heavy irrigation. Irrigation is given in most places once in seven to ten days. Stagnation of water in the soils is not very congenial to the proper growth of banana and, hence, the drainage of soil is also essential.

Harvesting

Early varieties commence flowering in southern and western India about seven months after planting, and the fruits take about three months more to ripen. In the Andhra Pradesh delta areas, the fruits are ready for harvesting about seven to eight months after planting. The first crop of the 'Poovan' variety matures in 12 to 14 months and the second in 21 to 24 months after planting. In other parts of India, the first crop is usually gathered a year after planting, whereas the succeeding crop may be ready in six to ten months thereafter.

The bunch is harvested just before it attains the ripening stage. When the fruits have reached the full size, they become plump, and mature with a distinct change in colour. For long transport, the bunch may be harvested somewhat earlier. The bunch is cut, retaining about 15 cm of the stem above the first hand. The yield varies considerably from 26,000 to 55,000 kg per hectare.

Curing and Marketing

The ripening of banana is done in several ways, *e.g.* exposing the bunches to the sun, placing them over a hearth, wrapping them in closed godowns or smoking them in various ways. One of the common ways is to heap the fruits in a room and cover them with leaves, after which fire is lit in a corner and the room is closed and made as air-tight as possible. Ripening takes place usually in 30 to 48 hours. In a cool store, the bunches ripen well at about 15° to 20°C. The application of vaseline, a layer of clay or coal-tar to the cut-ends of the stalks prevents rotting during ripening and storage. Wrapping up the fruits and packing them in crates help to reduce the damage during transport.

MANGO

Mango (Mangifera indica L.) occupies nearly half of the total area under fruits in the country. It is adaptable to a wide range of soil and climatic conditions and grows well right from Assam to the southern-most limits of the country and from the sea-level up to about 1,500 metres.

Fig. Mango

It withstands both fairly dry conditions and heavy rainfall, provided severe and recurring frosts in winter do not endanger the young trees.

Varieties

The number of varieties is very large. Each variety has its own peculiar taste, flavour and consistently of pulp. Some of the important commercial varieties grown in different regions are: 'Bombay yellow', 'Alphonso', 'Gopal Bhog', 'Zafran' (all early), 'Langra', 'Desheri', 'Safeda Lucknow', 'Safeda Malihabad', 'Fajrizafrani' (all mid-late). 'Fajri', 'Same Bihisht', 'Chausa', 'Taimura' (all late). In Uttar Pradesh; 'Bombai', 'Alphonso', 'Hemsagar', 'Krishna Bhog', 'Aman Dasheri', 'Gulab Khas' (all early), 'Langra', 'Aman Abbasi', 'Khasul-Khas' (all mid-late), ' Sinduri', 'Sukal', 'Taimuria' (all late) in Bihar; 'Bombai' or 'Maldah', 'Gopal Bhog', 'Hemsagar' (all early), 'Krishna Bhog', 'Zardalu' (both mid-late), 'Murshidabadi', 'Fazli Maldah' (both late) in West Bengal; 'Alphonso', 'Pairi', 'Cowsji Patel', 'Jamadar' in Bombay; 'Swarnarekha', 'Benishan', 'Cherukurasan', 'Panchadarkalasa', 'Desavathiyamamidi', 'Sannakulu', 'Nagulapalli', 'Irsala' in Circars; 'Rumani', 'Neelum Benishan', 'Bangalore', 'Alampur Benishan' in Rayalaseema; 'Murshidabadi', 'Mulgoa', 'Goabunder', 'Benishan', 'Neelam', 'Totapuri' or 'Bangalora' in Telengana; 'Alphonso', 'Peter', 'Rumani' in central districts; 'Mundappa', 'Neelam', 'Alphonso', 'Olour', 'Benn*et Al*phonso', 'Kalepad', 'Peter', 'Fernandin' in Coorg and Karnataka; and 'Padiri', 'Alphonso', 'Peter', 'Neelum', 'Bangalore', 'Rumani' in Tamil Nadu. In Goa, some excellent varieties like 'Alphonso', 'Fernandin', 'Mankurad' and 'Moussorate' are under cultivation. The new mango variety, 'Mallika' evolved at the Indian Agricultural Research Institute is now gaining popularity.

Other varieties, such as 'Jehangir' and 'Himayuddin', produce high-quality fruits, but are poor in yield and cropping tendencies. Attempts are being made to evolve hybrid progenies by crossing.

Propogation and Planting

Propogated vegetatively by inarching or budding in situ in the nursery, either by using Forkert or by using the T-method. The beginning of the monsoon in light-rainfall areas and the end of the monsoon in heavy-rainfall regions are the most suitable periods for inarching or budding. Recently, veneer-grafting has been found to be the best method of mango propagation.Grafted plant are ready for transplanting in the field after six to twelve months. Select straight-growing grafts and set them in pits filled with soil mixed with farmyard manure (45 kg) and a fertilizer mixture containing 0.225 kg of N, 0.45 kg of P and 0.225 kg of K per pit. The planting-distance is 7.5 to 9 metres in poor shallow soils and 15 to 17 metres in deep fertile soils.

The beginning of the monsoon in low rainfall areas or the end of the monsoon in heavy rainfall tracts is the best time for planting. The graft-joint should be at least 15 cm above the ground.

Pruning

No systematic pruning is done. The removal of dead-wood and the thinning of over-crowded and mis-shapen branches after about four years is all that is necessary; flowers that appear during the first three or four years should be removed.

Culture

Before planting, the field is ploughed, harrowed and levelled. Thereafter, it is ploughed and harrowed twice a year, once in the beginning of the monsoon and again at the close of the rainy season or in the cold-weather. It is green-manured once every two or three years.

Short-season intercrops, like vegetables, may be taken during the first four to five years. Young plants require irrigation regularly. After five to six years, when they have established themselves, the trees are able to grow and fruit satisfactorily without irrigation in most parts of Peninsular India. In northern India, they have to be irrigated throughout their life. Irrigation is usually withheld during the cold weather before flowering, especially in deep retentive soils.

Though the exact manurial requirement is not known, regular manuring is beneficial. The dose recommended for the bearing trees is 45 to 70 kg of farmyard manure, 0.5 to 0.7 kg of N, 0.7 kg to 1.0 kg of P and 1.2 to 1.5 kg of K per tree. Nitrogen and half of potash may be given before the monsoon, and farmyard manure, phosphate and half of potash in October or before flowering starts.

Crop Irregularity

Grafted mango-trees bear fruits from the fourth or fifth year onwards and a full crop from the tenth or fifteenth year. The erratic bearing of mango is well known. It depends upon the variety, the weather and climatic conditions and cultural treatments.

The selection of regular-bearing varieties, timely cultural practices and proper nutrition help to produce a regular crop. New growth in spring, on which flower-buds are produced during the next winter, can be encouraged by applying nitrogenous fertilizers (0.45 to 0.90 kg of N per tree).

In the case of heavy late rains, an additional ploughing in winter helps to produce flower-buds in January-February. In the case of individual trees, ringing or girdling in August-September may also to help to force flower-buds the following winter. The application of Ethral (200 ppm) from September onwards has been found to induce flowering in mango in Karnataka by the Indian Institute of Horticultural Research.

Improvement of Old and Seedling-Trees

Mango-trees of inferior varieties, so also those raised from seeddlings, can be converted into choice varieties by grafting them in situ either by crown or side-grafting. In crown-grafting, the trunk of the tree is cut down to about half a metre from the ground and one or more scions of the selected variety are inserted into it between the bark and the wood by splitting open the bark. The scion should be a dormant, terminal shoot, about 12.5 mm in diameter, with a whorl of plump swollen buds at the top. In side-grafting, the procedure is the same as in crown-grafting, except that the trunk of the stock tree above the grafting joint is cut down after the scions have sprouted and have established themselves properly. Old trees, having several branches, can be similarly improved (top-worked) by crown-grafting on each branch at a suitable height. Sometimes, the grafting is done by inarching, but the process is cumbersome, expensive and not very satisfactory.

Harvesting and Marketing

The fruit takes five to six months to mature. Depending upon the onset of flowering, the mature fruits are ready for harvesting from April to May in western India, from May to June in the Deccan, from February to March in Malabar, from April to July in the coastal Andhra Pradesh, from May to August in Mysore and Rayalaseema, and from June to August in northern India. The mature fruits are harvested by severing the stalks to which they are attached, when they are still green and hard.

The signs of maturity vary with different varieties. As a mango tree usually bears flowers in three or four distinct flushes lasting over a month, it is preferable to harvest the fruits as they mature. The fruits, so harvested, can be transported after packing them in baskets or wooden crates, properly padded with straw, wood-shavings or wool, to long distances. For overseas markets, they are packed in a single layer in specially designed wooden crates.

For ripening, the fruits are spread out on rice straw in a single layer. Two or three such layers are built one above another in a well-ventilated room. The mangoes are ready for disposal after they change colour.

Yield varies considerably with the variety, vigour of growth, flowering, etc. A grafted tree yields about 300 to 500 fruits in the tenth year, about 1,000 in the 15th year and 2,000 to 5,000 from the 20th year onwards.

TRENDS IN HORTICULTURE GROWTH

Tamil Nadu is one of the leading horticulture States in India contributing 7.7 per cent to the National Horticultural production with 5.7 per cent of the national level area. Tamil Nadu has been blessed with diversified agro-climatic conditions, suitable for a wide range of horticulture crops like fruits, vegetables, spices, plantation crops, flowers and medicinal plants.

A large extent of wastelands and under-utilized lands are available in the State for horticulture development. Tamil Nadu has a long coastal belt of 1000 km. suitable for crops like cashew, coconut, tropical orchids etc. The southern part of Tamil Nadu has the potential for growing off-season mangoes and grapes. A lot of awareness has been created among the farmers of Tamil Nadu about cultivation of high value horticulture crops. It is aimed to achieve 8 per cent annual growth rate during X Five-Year Plan in the horticultural sector. The area coverage in 2003-04 was 8.24 lakh hectares, a little less than the normal area of 8.52 lakh hectares. During 2004-05, despite the failure of the monsoon, the area coverage would still be around 8.91lakh hectares. In the year 2005-06, an area of 9.73 lakh hectares is expected to be covered.

TAMIL NADU HORTICULTURE DEVELOPMENT MISSION

The Government have setup a Mission for Horticulture development in Tamil Nadu vide G.O. Ms. No.155, Agriculture (H1-2) Department, dated 28.4.2003 with a mandate to give impetus to production, processing for value addition and marketing of Vegetables, Fruits, Flowers and Medicinal Plants. Tamil Nadu is the first state in India to setup a separate Mission for development of Horticulture in the year 2003. It is aimed to double the production of horticulture crops by the year 2011-12 in Tamil Nadu. The present level of productin of 99.47 L.MT will be doubled to 187.80 L.MTs during the year 2011-12.

OBJECTIVES OF THE MISSION

- Improving production through balanced nutrition management.
- Evolving suitable mechanism for regulating the production of quality planting materials and giving impetus to need based research.
- Establishing adequate infrastructure for post harvest management especially preservation and marketing.
- Encouraging active involvement of farmers associations in adoption of modern technological practices.

A Governing Body has been constituted to monitor the effective implementation of the mission under the Chairmanship of the Honourable Minister for Agriculture with the Vice- Chancellor of Tamil Nadu Agricultural University, Secretaries to Government and Senior officers of various departments and Nominees of Central Government agencies like National Horticulture Board, APEDA (Agricultural and Processed Food Products Export Development Authority) as members. The Chief Secretary to Government is the Vice Chairperson of the Governing body.

Mango Development Mission

Under Mango Development Mission it has been programmed to layout demonstration plots in the farmer's field to enable them to have a visual impact of various new technologies involved in the cultivation of Mango crop. A sum of Rs.22.55 lakhs has been proposed for this purpose.

National Horticulture Board has agreed to sanction Rs. 10.00 lakhs out of which released a sum of Rs.5.00 lakh for implementation of this scheme. On receipt of Government Order, this scheme will be implemented in the selected Mango growing districts.

Cashew Development Mission

Under Cashew Development Mission it has been programmed to layout demonstration plots in the farmer field to increase the productivity of cashew with high yielding varieties and hi technologies. A sum of Rs.23.00 lakhs has been proposed for this scheme. National Horticulture Board has agreed to sanction Rs.10.00 lakhs out of which released a sum of Rs. 5.00 lakh for implementation of this scheme. On receipt of Government Order this scheme will be implemented in the selected cashew growing districts.

Training to Farmers

With a view to expose the farmers to various hitech aspects of Horticulture, farmers are taken to various research stations in out of state on tour-cum-training basis. About 10,000 Nos. of farmers are given training during the year 2004-2005. Apart from this farmers were given awareness in various horticulture technologies of Medicinal Plants, Flower cultivation and Vegetable Cultivation through the conduct of District level seminars and Exhibition.

The extension officers and field level workers of this department are also given training in the latest technologies of horticulture from the Research Stations situated inside and outside the State.

STATE PLAN SCHEMES

The following State Plan Schemes have been implemented during 2004-05.

Integrated Horticulture Development Scheme (IHDS)

In G.O. Ms. No.180, Agriculture (H2) Department, dated 28.5.2004, Integrated Horticulture Development Scheme was sanctioned with an outlay of Rs.766.26 lakhs. Under this scheme, planting materials, high yielding/hybrid vegetable seeds are distributed to horticultural farmers at 50% subsidised cost.

Integrated Tribal Development Programme (ITDP)

In G.O. Ms. No.189, Agriculture (H2) Department, dated 28.7.2004, Integrated Tribal Development Programme was sanctioned with an outlay of Rs.39.325 lakhs. Under this scheme, horticulture inputs are distributed to Tribal farmers in Salem, Namakkal, Dharmapuri, Tiruvannamalai, Vellore, Trichy and Villupuram districts at subsidised cost. Tour-cum-Training is also given to the Tribal farmers under this scheme.

Western Ghats Development Programme (WGDP)

In G.O. Ms. No.52, Plannting, Development and Special Initiatives (TC-1) Department, dated 22.7.2004, Western Ghat Development Programme was sanctioned. Under thisscheme, horticulture inputs are distributed to the horticultural farmers in Theni, Madurai, Dindigul, Virudhunagar, Coimbatore, Erode, Tirunelveli and Kanyakumari districts at 25% subsidised cost.

Hill Area Development Programme (HADP)

In G.O. Ms. No.60, Planning, Development and Special Initiatives (TC 1 & 2) Department, dated 5.8.2004, Hill Area Development Programme was sanctioned with an outlay of Rs.315.00 lakhs.

Under this scheme, horticultural inputs are distributed to the farmers of Nilgiris district at 25% subsidised cost.

The above Schemes will be continued during the year 2005-06 also. In the above schemes:-

- Focus is given for cultivation of high yielding / hybrids and varieties, adoption of efficient water and nutrient management and Integrated Pest Management.
- Adoption of location specific and crop specific technologies is encouraged.
- Mango, Amla, Guava, Sapota and Vanilla are given importance under fruits and spices categories. High yielding / export potential varieties are given priority.
- Under vegetables, focus is given on hybrids / high yielding varieties which fetch good return to the farming community.
- Under Integrated Horticulture Development Scheme, horticulture plants are distributed at 50% subsidy for a maximum of 1 hectare for fruits and 0.5 hectare for seeds of vegetables and flowers to each beneficiary.
- In the Tribal areas of Tamil Nadu, *viz.*, Shevroy hills, Arunoothumalai, Pachamalai, Kolli hills, Sitheri hills, Javvadu hills, Kalrayan hills, and the tribal farmers are encouraged to develop individual orchards by optimally utilizing the available water resources.
- Considering the economic status of the tribal farmers, the horticulture plants are distributed at 75% subsidy under the Integrated Tribal Development Programme (ITDP).
- Watershed approach is followed in the Western Ghat Development Programme and the Hill Area Development Programme.
- Focus is given on prevention of soil erosion and preserving the Eco-System through appropriate cropping pattern.
- Planting materials and seeds are distributed at 50% cost under Western Ghat Development Programme and Hill Area Development Programme.
- The State is maintaining 55 State Horticulture Farms in which Pedigree and Hybrid planting materials are produced and supplied to the farmers under various schemes.

- To update the technical knowledge of the extension officers of the Horticulture Department, in-service training is given at Central Horticulture Training Centre, Kudumianmalai in Pudukkottai district.
- A two-year diploma course is being conducted at the Horticulture Training Centre, Madhavaram (Chennai) with an intake of 40 students every year.

Promotion of Alternate Cropping in the Nilgiris

Promoting alternate crops is considered as an important measure to overcome the crisis facing the tea growers. Steps are being taken to popularize multi-tier cropping, *viz.* cultivation of silver oak, pepper, cardamom, mandarin orange, etc., along with tea plantation. Considering the prevailing agro-climatic condition in the Nilgiris, new kinds of crops like Macadamia and Peacan nuts have been identified for introduction.

It is proposed to raise these crops in the State Horticultural Farms at Nanjanad and Colegrain in Nilgiris district on trial basis. Subsequently, efforts would be made to introduce these crops to the farmers in select pockets.

A sum of Rs.50.00 lakhs has been sanctioned vide G.O. Ms. No.244, Agriculture (H2) Department, dated 17.6.2004 by the Government for promoting alternate cropping in Nilgiris district and the scheme will be continued during 2005-06 also.

Tamil Nadu Precision Farming Project

As announced during last budget session, Government have sanctioned an Innovative Scheme by name "Tamil Nadu Precision Farming Project" to be implemented in Dharmapuri District and newly formed Krishnagiri district covering an area of 400 Ha. of farmers' land with Micro Irrigation and Fertigation vide G.O. Ms. No.24, Agriculture (H1) Department, dated 23.1.2004. This scheme will be implemented with a total cost of Rs.720.60 lakhs over a period of 3 years. It is programmed to cover an area of 100 Ha. in the first year, 200 Ha. in the second year and 100 Ha. in the third year at financial allocation of Rs.208.59 lakhs, 316.16 lakhs and Rs.195.85 lakhs respectively.

The Tamil Nadu Agricultural University will undertake this project as a Turnkey project and implement it with the co-operation of the Departments of Horticulture, Agricultural Engineering, Agriculture, Agricultural Marketing and Agri-Business and the District Administration. High value like gherkins, hybrid tomatoes, capsicum, paprika, babycorn, white onion, bhendi, cabbage and cauliflower are proposed to be cultivated under the scheme. Under this project, 100% subsidy will be given on the cost of cultivation of the first crop to the farmers selected during first year. 10% of the cost of cultivation will be collected from the farmers selected during second year. 20% of the cost of cultivation will be collected from the farmers selected during third year. During the year 2005-06, the above scheme will be implemented in six more districts *viz.*, Theni,

Vellore, Cuddalore, Erode, Tirunelveli and Thanjavur with an financial outlay of Rs.10.00 crores. An area of 100 Ha. will be covered in each district.

SCHEMES

Under Part II Scheme a sum of Rs.120.00 lakh was sanctioned vide G.O. Ms. No.398, Agriculture (H1) Department, dated 6.8.2004 to improve the irrigation facilities in 28 State Horticulture Farms and the work is entrusted with Public Work Department and Agriculture Engineering Department and will be completed before 31.3.2005.

Tamil Nadu Horticulture Development Agency (TANHODA)

An independent nodal agency *viz.*, The Tamil Nadu Horticulture Development Agency (TANHODA) has been formed vide G.O. Ms. No.250, Agricultue (H1) Department, dated 18.6.2004 and Rs.20.00 lakhs was sanctioned under Part II 2004-05 with the following objectives.

- To promote Hi-Tech Horticulture and Precision Farming.
- To promote Public Private Partnership (PPP) for horticulture development.
- To disseminate knowledge on modern technologies in horticulture to the farming community as well as extension personnel.
- To manage production and distribution of pedigree planting materials through State Horticulture Farms.
- To implement the horticulture development schemes including the schemes under the Tamil Nadu Horticulture Development Mission and the National Horticulture Mission by accessing funds from the agencies concerned.

This agency will serve as a special purpose vehicle (SPV) for the implementation of various horticultural development programmes.

Part II Schemes (2005-06)

The following new schemes will be implemented during the year 2005-06.

- Scheme for provision of infrastructure facilities in 25 State Horticulture Farms (Farm equipments like Power Tillers with Accessories and Multipurpose Tanker) - Rs.50.00 lakhs.
- Organic farming vegetable cultivation with market linkage in 15 State Horticulture Farms - Rs.10.00 lakhs.
- Water Management - Command Area Development Programmes (Centrally sponsored scheme - shared between State and Centre)

Command Area Development Programme

The Command Area Development Programme is implemented by the Agricultural Engineering Department as a centrally sponsored and equally shared programme on 50: 50 basis between the State Government and

Government of India. At present, the programme is implemented in Cauvery Command, Parambikulam Aliyar Project, Tambirabarani River Basin Project, Palar-Poranthalar Project, Krishnagiri Reservoir Project, Gadana Ramanadhi Irrigation System, Nambiyar River Basin System, Patchaiyar River Basin System, Manimuthar Irrigation System, Aanaimaduvu System, Chinnar Resevoir system and Maruthanadi System.

The programme has been completed in Lower Bhavani Project, Cumbum Valley Project, Periyar - Vaigai Project, Sathanur Reservoir Project, Amaravathy Reservoir Project, Kothayar-Chittar project and Thoppaiyar Project.The remaining projects will be covered as per the schedule approved by the Government of India. The six major components of Command Area Development Programme are as follows:

- Construction of Field Channel
- Construction of Field drain.
- Reclamation of Water logged Saline and alkaline soil areas.
- Rotational Water Supply.
- Participatory Irrigation Management
- Management Subsidy.

The Agricultural Engineering Department has also developed a water management strategy for the area below the sluice, which is crucial for agricultural production. Water management holds the key for optimisation of water use and maximisation of crop production.

Construction of Field Channel

Construction of Field channel is done by considering the needs of the head reach farmers so that their interference is eliminated and the tail end farmers also get adequate irrigation water.

This brings in equity in distribution of water to all the farmers in the entire sluice command. This has been achieved by using optimum flow concept.

The unit cost is Rs.10000/- per ha. in all New Project areas including Cauvery Command. For old ayacut of Cauvery Command the unit cost of construction of field channel works is Rs.4400/- per ha.

An area of 9.87 lakh ha. has been covered till 31.3.2004 out of total cultivable command area of 10.97 lakh ha. with Construction of field channel component in 12 command areas. During 2004-2005, this work is being carried out in an area of 25695 ha. It is programmed to continue the scheme during 2005-2006 also.

Construction of Field Drains

Under the construction of field drains component, drains from individual fields to Government drains (or) natural drains outside the outlet are executed.

The expenditure on field drains is inclusive of costs of earthwork, road cuttings and drop structures, if needed to prevent erosion. The unit cost per ha. for this component is Rs.4000/-.

Reclamation of Water Logged, Saline & Alkaline Soil Areas

In this component, the area affected by Waterlogging, saline and alkaline problems within the command area are identified and measures to reclaim those patches are takenup. At present this component is taken up in Tamirabarani River Basin Project and Krishnagiri Reservoir Project at an unit cost of Rs.15000/- per ha.

Rotational Water Supply Works

Rotational Water Supply or Warabandi is a system of equitable water distribution, by turns according to a predetermined schedule specifying the day, time and duration of supply to each farmer in proportion to his holding size in an outlet command.

The rotational water supply schedule is prepared by the staff of Agricultural Engineering Department after executing the on farm development works and handed over to farmers for implementation. Rotational water supply was first introduced in Lower Bhavani Project, Periyar - Vaigai Project and Cauvery Command in the year 1984-'85. An area of 8.92 lakh ha. has been covered under rotational water supply in Tamil Nadu in 12command areas till March 2004.

During 2004-2005, this work is being taken up in an area of 52750 ha. It is programmed to continue the scheme during 2005-2006 also.

Participatory Irrigation Management

To implement On Farm Development Programmes and for ensuing equity in distribution of water, farmers participation is important.

Agricultural Engineering Department is intensively encouraging itself in this work for the last 20 years. The task of handing over the responsibilities to the farmers is being implemented. In orders to ensure farmers participation, farmers associations are formed on a three tier basis as follows.

Farmers' Association - at sluice command level
Farmers' Council - at distributory level
Farmers' Federation - at project level.

Management Subsidy

In order to sustain the farmers associations/ councils already formed and for the maintenance of the on-farm infrastructures created and to continue RWS (Warabandhi) for equitable water distribution, a financial assistance is provided in the form of one time functional grant to registered Farmers' Councils at Rs.540.00 per ha. shared equally by the Central and the State Governments. To avail this, farmers have to contribute Rs.60.00 per ha.

Out of the 632 Farmer's Councils registered, an amount of Rs.1478.95 lakhs has been released as management subsidy till March 2004 to 597 registered Farmers' Councils to maintain the infrastructure created under the Command Area Development Programme.

WORLD BANK AIDED MODELREHABILITATION PROJECT IN HANUMANADHI SUB BASIN

The World Bank aided IAIP pilot project was implemented in Hanumanadhi sub basin covering 135 ha. in the command area, which is fed by Adavinainarkoil system tanks, at a cost of Rs.81.00 lakhs.

This project is construed as a threshold for a broader Model Rehabilitation Project in Hanumanadhi sub basin.

The World Bank aided model Rehabilitation Project in Hanumanadhi sub basin of Tamirabarani river basin in Tirunelveli district has been successfully completed by September 30th 2004.

The project has been executed and implemented in a multi disciplinary project approach involving PWD-WRO, Departments of Agricultural Engineering, Agriculture, Horticulture, Fisheries, Agricultural Marketing and TNAU.

The funds of Rs.62.5 Crores was provided vide GO MS No. 33 PW (WR 1), Dated 30-1-2004, out of which Rs. 22.55 Crores was allotted to Agricultural Engineering Department. The project area encompasses Sengottai, Tenkasi and V.K. Pudur taluks.

The ayacut irrigated by 50 tanks and 14 anicuts isspread over an area of 3000 ha. in 15 villages namely Ayikudi, Elathur, Kilankadu, Kodikurichi, Kulayaneri, Kutthukalvalasai, Nainaragaram, Neduvayal, Panpoli, Rajagopalaperi,Sambavar Vadagarai, Surandai, Sundarapandiyapuram, Verakeralampudur and Vadagaraikilpidagai..An area of 2194 Ha. has been covered with an expenditure of Rs. 16.35 crores by the Agricultural Engineering Department.

Objectives

- Improving productivity of water and crop production by sharing of water and Mechanized farming.
- Uniform and optimum application of water and fertilizer by high tech micro irrigation systems. Rain water harvesting and recharge of ground water through rejuvenation of wells and farm ponds.
- Training, effecting attitudinal changes among the stakeholders and improving their socio - economic status with measures like Aquaculture in farm ponds,

High Tech Drip and Sprinkler facility

In the pre project period only two crops were cultivated in a year and most of lands were kept fallow for want of water in the third season After implementation of the project it is made possible to cultivate an additional third crop using the water conserved through micro irrigation techniques. After the project implementation an intensive cropping scheme one like below is followed.

- Crop : 28 Ac-inch water required crops like groundnut, vegetables.
- Crop : Paddy.
- Crop: 18 Ac-inch water requirement crops like Pulses, fodder cholam etc.

The introduction of drip and sprinkler irrigation systems in the project area on a 10 Ha block basis is estimated to save atleast 30 - 40 % total irrigation water.

Sprinkler irrigation system are provided to the individual farmers who do not have wells and new borewells are sunk for these farmers.

Drip irrigation systems are provided to the farmers who have atleast one well.

Under ground PVC pipes are laid to convey water to the micro irrigation systems from the sources of water. However open field channels shall carry water during the paddy season.

An area of 2194 Ha., is covered with an expenditure of Rs.1065.72 crores under high tech micro irrigation system including pipe laying.

GIS and Survey

The geographical information survey is used to get precise information on land holding, water bodies, vegetation etc., The GIS format creation in 1:1000 scale, high resolution satellite data like IRS PAN merged 1:10,000 scale, GIS in 3-D, aerial photographs are some of the advanced survey and map preparation techniques.

ISDN / Lease line connections are installed at offices of water users groups, AED office in Tenkasi, Tirunelveli and MDPPP office AED, Commissioner of Agricultural and Commissioner of Agricultural marketing, secretariat in Chennai to provide latest information on farm practices and market trends to the farmers. These facilities have been created with Rs. 2.00 crores.

Water Harvesting Structures

Farm ponds are executed for collection of rainwater. 52 ponds have been constructed.

The pond water is used for domestic, irrigation and aquaculture purposes. The defunct and low yielding wells are rejuvenated with vertical and horizontal in - well bores to improve their yield.

Shallow and medium bore wells are sunk and the new bore wells are provided with energized motors. An amount of Rs. 2.84 crores has been incurred on these water-harvesting measures.

Farm Mechanization

Modern agricultural machineries like power tillers, paddy transplanters etc. have been procured at the cost of Rs.0.53 crores and are hired to the farmers.

Training to Farmers and Officers

Trainings on high tech micro irrigation systems, water-harvesting methods and farm mechanization are imparted to the water user groups and departmental officers with Rs. 0.34 crores. Documentation, video graphy and evaluation studies of the project are also included in the programme.

Farmers' Participation

The participation of the stakeholders - the farmers, is emphasized right from the decision-making and implementation stages. The farmers contributed in cash, kind and labour at various stages of the project. The progressive farmers, the women farmers and rural youth participated in every sphere of activity. The farmers will be sharing a portion of the project cost in the form of betterment levy.

Follow up on Project

With the funds from the ongoing schemes, state and other organizations like NABARD etc., the uncovered areas in this project are proposed to be covered.

JATROPHA

It has been proposed to taken up Jatropha Cultivation in 200 Ha. funded by NOVOD Board during the year 2004-05 in the following 11 State Horticulture Farms.

- G.O. Vallathirakottai, Pudukottai District 45 Ha.
- G.O. Karumanthurai, Salem District 50 Ha.
- SHF, Jeenur, Krishnagiri District 20 Ha
- SHF, Kudumianmalai, Pudukottai District 10 Ha
- SHF, Navlock, Vellore District 20 Ha
- SHF, Mulluvadi, Salem District 10 Ha
- G.O. Srivilliputhur, Virudhunagar District 12 Ha.
- SHF, Melkadirpur, Kancheepuram District 10 Ha
- SHF, Vichanthangal, Kancheepuram District 10 Ha
- SHF, Thagarakuppam, Vellore District 10 Ha
- SHF, Kannampalayam, Coimbatore District 5 Ha

Total 200 ha.

NOVOD has sanctioned a sum of Rs.75.00 lakhs to the Managing Director, Tamil Nadu Watershed Development Agency, Chennai 32 towards implementation of this programme.

TARGETS AND ACHIEVEMENTS

The financial allocation for the year 2004-05 and the expenditure incurred upto February 2005 and theproposed outlay for the year 2005-2006 for the State Plan Schemes and Schemes shared between State and Centre are furnished below.

Table. Allocation for the year 2004-2005 and Proposed outlay for 2005-2006.

Unit: Rs. in lakhs

Sl. Scheme Amount Sanctioned	2004 – 2005 Acht. upto Feb. 05	2005-2006 Proposed Outlay	No
A. State Plan Schemes			
1 Integrated Horticulture Dev. Scheme	766.260	615.080	769.400
2 Integrated Tribal Dev. Programme	39.330	35.135	41.790
3 Western Ghat Dev. Programme	135.480	105.158	132.080
4 Hill Area Development Programme	315.000	267.750	314.940
5 Precision Farming in Dharmapuri Dt.	232.950	203.000	1120.850
6 Promotion of alternate cropping(Nilgiris Dt)	50.000	48.090	50.000
7 Part-II Schemes	120.000	105.000	60.010
8 TANHODA	20.000	0	20.000
Total	**1679.020**	**1379.213**	**2509.0 0**
B. Schemes shared between State and Centre			
1 Integrated Prog. for Dev. of Cashew	155.200	141.847	273.320
2 Integrated Prog. for Dev. of Fruits	218.000	186.389	318.350
3 Integrated Prog. for Dev. of Spices	60.000	47.946	180.130
4 Integrated Prog. for Dev. of Vegetables	219.000	191.644	310.000
5 Integrated Prog. for Development of Medicinal and Aromatic Plants	52.500	48.350	52.700
6 Integrated Prog. for Dev. Cocoa	12.000	11.190	12.000
7 Dev. of Commercial Floriculture	145.820	111.200	165.000
8 Development of Mushroom cultivation	10.000	9.500	10.000
9 Development of Horticulture through plasticulture interventions	14.500	13.030	14.500
10 Innovative Programmes for Development of Horticulture	113.900	43.455	90.000
Total	**1000.920**	**804.471**	**1426.000**
Grand Total (A+B)	**2679.940**	**2183.684**	**3935.070**

HORTICULTURE AND ANTHROPOLOGY

The origins of horticulture lie in the transition of human communities from nomadic hunter gatherers to sedentary or semi-sedentary horticultural communities, cultivating a variety of crops on a small scale around their dwellings or in specialized plots at some remove (such as the "milpa" or maize field of mesoamerican cultures). In forest areas such horticulture is often carried out in swiddens ("slash and burn" areas). A characteristic of horticultural communities is that useful trees are often to be found planted around communities or specially retained from the natural ecosytem. Horticultural communities may be distinguished from agricultural ones by:

- The small scale of the cultivation, using small plots of mixed crops rather than large field of single crops
- The use of a variety of crops, often including fruit trees
- The encouragement of useful native plants alongside direct cultivation
- Continued use of other forms of livelihood.

In pre-contact North America the semi-sedentary horticultural communities of the eastern woodlands (growing maize, squash and sunflower) contrasted markedly with the mobile hunter gatherer communities of the Plains people.

In central America, Mayan horticulture involved augmentation of the forest with useful trees such as papaya, avocado, cacao, ceiba and sapodilla.

In the cornfields, multiple crops were grown such as beans (using cornstalks as supports), squash, pumpkins and chili peppers, in some cultures tended mainly or exclusively by women.

PESTS AND DISEASES OF ROSES

Roses Rosa sp. are susceptible to a number of pests, diseases and disorders. A large number of the problems affecting roses are seasonal and climatic. Certain varieties of roses are naturally more resistant or immune than others to certain pests and diseases.

Cultivation requirements of individual rose species and cultivars, when observed, often assist in the prevention of certain pests, diseases and disorders.

Pests

Pests are often considered to be the insects that affect roses:

- *Aphids (Greenfly) (Order Hemiptera: Family Aphididae) Macrosiphum Rosae:* Likely to be found on new shoots and buds, aphids are soft bodied insects 1-2mm long. Often green but occasionally light-brown, and sometimes with wings, they may cover (in a colony) the complete growing tip of the plant. Aphids are most active in spring and summer and multiply at a prodigeous rate feeding on the sap of the plant by piercing the plant cells via a proboscis. In large quantities they may seriously retard the growth of the plant and ruin buds. They are particularly damaging to the new shoots with subsequent damage to the emerging leaves which become malformed with much the same appearance as leaf-curl in peaches.

- *Two-spotted Mite (Spider-mites or Red spider mite) (Order Acari: Family Tetranychidae) Tetranychus Urticae:* Previously known as red-spider mite these arachnids prefer the underside of leaves and are difficult to seewith an unaided eye. Evidence of their presence is silvering of leaveswhere the mites have destroyed individual leaf cells. Fine webbing and eggs on the undersides of leaves is further evidence of the presence of Tetranychus urticae.
- *Thrips (Order Thysanoptera):* Thrips are slim-winged insects 1mm in length, resembling fine black slivers of wood. Preferring light-coloured blooms and often appearing in plague numbers flowers are often left looking bruised and lustreless.
- *Caterpillars (Order Lepidoptera) See also List of Lepidoptera which Feed on Roses:* The tortryx (tortrix) moth Lozotaenia forsterana is a prominent pest of roses, although not the sole pest. The caterpillars are green, up to 15mm long, and can be found boring into buds or within curled leaves. When disturbed the caterpillars move swiftly, dropping to the ground on a fine thread. Damage is chewn leaves and flowers and buds with "shot holes".

Cottony Cushion Scale (Order Hemiptera : Family Coccoidea) Icerya purchasi

This scale infests twigs and branches. The mature female is oval in shape, reddish-brown with black hairs, 5 mm long.

When mature the insect remains stationary and produces an egg sac in grooves, by extrusion, in the body which encases hundreds of red eggs. The insect causes little damage but produces copious honeydew (frass) that can cause damaging sooty mould.

California Red Scale (Order Hemiptera : Family Coccoidea) Aonidiella Aurantii

A hard scale, orange to orange-pink, the female covering being less than 1.5mm across. Often in plague numbers this scale infests upper surfaces of foliage causing yellowing, leaf fall, and twig and branch dieback. Serious infestations can cause plant death.

Rose Scale (Order Hemiptera : Family Coccoidea) Aulacaspis rosae

Mainly found on the stems and branches of the plant, lack of control will allow the pest to spread to flower stalks and petioles.

At this point the plant would be stunted, spindly and with a white, flaky crust of scales on the bark. Female Aulacaspis rosaemay live for 1 year and may lay 80 eggs each with several overlapping generations living within milliimetres of the original parent.

Leaf cutting bee (Order Hymenoptera : Family Megachilidae) Megachile spp.

Leafcutter bees are 6-16mm long and mostly black with bands of light-coloured hair. They chew pieces from the edges of leaves. The pieces are regular in shape, circular or oval. Damage is not often significant.

Nematodes (Eelworms)(Order Tylenchida: Family Heteroderidae)

Root-knot nematode Meloidogyne spp.

See - Root-knot nematode - symptoms of Meloidogyne infestation in roses is stunting, slow-growth, pale green leaves and wilting in mild weather.

Metallic flea-beetles (Order Coleoptera: Family Chrysomelidae) Altica spp.

The small, shiny and metallic Altica beetles have thickened hindlegs adapted to jumping, similar to fleas. The insects are 3mm long and chew holes of irregular shapes in young leaves and buds. As the leaves enlarge so do the holes.

Diseases

Fungal, bacterial and viral diseases affect roses:

Fungal Disease

- *Black Spot (Class Leotiomycetes: Family Helotiales) Marssonina rosae syn. Diplocarpon rosae :* Marssonina rosae causes black spots on leaves. The spots, which may be as much as 12mm across, are generally circular and have an irregular edge often with a yellow halo. Leaves frequently turn yellow and fall early. Sometimes new leaves are produced, and these may also become affected.
  Continual defoliation will cause weakness, die-back or death of the plant. Some very susceptible species may have stems affected with a considerable reduction in plant vigour.
- *Powdery Mildew Oidium sp. :* Oidium produces a very fine, powdery coating on the surface of buds and leaves. Significant cases have stems and particularly thorns, infected. Attacks on young leaves and buds will cause deformity with retardation of growth. Infected buds will fail to open. The disease is likely in hot, humid weather, with fungal spores overwintering on the stems and fallen leaves.
- *Downy mildew (Class Oomycetes : Family Peronosporaceae) Peronospora sparsa :* Peronospora causes purple-red to dark-brown spots on the leaves with irregular margins, however, often angular. Stems, petioles and flower stalks can split and spotted with purple marks. Buds,

sepals, petals and calyces can be affected and will present purple spots. New growth affected will be deformed. The disease is spread by wind.

- *Rust Phragmidium mucronatum :* Rose rust appears as yellow patches on the surface of leaves, with orange pustules of spores underneath the leaf. The fungus is spread by wind. Affected leaves fall prior to healthy ones and plants may be defoliated in serious infections.
- *Anthracnose Sphaceloma rosarum :* Spots caused by this fungus originate from a point where leaves are watersoaked, ususally unnoticeable at first, until they turn black with a very distinct defined edge. As the spots enlarge the centre becomes gray and may fall out resulting in a shot-hole appearance. Defoliation may occur but is often not serious.
- *Grey mould (Class Leotiomycetes: Family Sclerotiniceae) Botrytis cineria:* On roses grey mould is prmiarily a disease of the flowers and buds, leaves are infrequently attacked. Infected buds rot on the stem and infection may progress down the stem. On petals botrytis cineria produces pink rings.
- *Verticillium wilt (Class Incertae sedis: Family Verticillium) Verticillium dahliae*
- *Sooty moulds Alternaria spp. :* Sooty mould appears as black, dry powder on leaves similar to chimney soot. Many sooty moulds grow on the honeydew (frass) produced by sap-sucking insect such as aphids and soft scales. Alternaria does no direct damage to plants but surface cover of leaves will reduce the plants capacity to photosynthesise and may create an unsatisfactory plant appearance.
- *Canker Leptosphaeria coniothyrium and Cryptosporella umbrina :* Cankers present as small yellowish or reddish spots on bark slowly increasing in size. Leptosphaeria coniothyrium turns brown, increases in size, and may eventually girdle the stem. The tissue within the infection begins to dry out and shrink, presenting a shrivelled appearance. If the disease infects only part of the stem, growth above the canker will continue. If it girdles the stem, however, growth will cease and the stem will die.

Viral Disease

- *Rose Mosaic :* This diesase is caused by a complex or viruses and is charaterised by yello patterns on the leaves. The patterns vary considerably, ranging between all-over fine blotches to patterns of lines in waves. The patterns may appear on a few or many leaves. Plants are infected by this virus at propagation using infected plant material.
- *Rose Wilt* : Rose wilt is a complex of viruses and is referred to as dieback in some areas. The disease can be spread by vectors such as

aphids. Symptoms are variable and range from stunted growth to curled young leaves. The soft tissue symptoms are more evident in spring and new leaves will reflex towards their own petioles. The affected leaves are brittle and easily fall from the plant. Fully formed leaves will 'wilt' as if the plant were water stressed.

Bacterial Disease

- *Crown gall rot (Class Alpha Proteobacteria: Family Rhizobiaceae) Agrobacterium rhizogenes :* This disease is characterised by large lumps at the base of the plant stem or on roots. Galls may appear higher on stems as the disease progresses. Galls are soft compared to surrounding plant tissues. The pathogenic bacteria enter the plant via a wound. If the disease affects the plant whilst it is young the plant may be affected to the degree where it will not produce blooms. All affected plants wilt readily and grow poorly.

Environmental Disorders

- *Frost* : Frost will destroy fresh growth causing stems and leaves to wilt, turn black and fall away from the plant. Timing pruning to promote growth after the threat of frost is a means to avoid frost damage.
- *Salinity* : Salinity will present in roses as limp and light brown leaves with dry leaf margins. Soil may require testing to determine salinity levels. Symptoms will present if salinity is greater than 1200 parts per million.
- *Herbicide Damage* : Overspray or soil leaching of herbicidal sprays can present with several symptoms: Prolonged exposure to overspray of glyphosatewill cause yellow leaves and new leaves will be small and elongated.

Hormone weedsprays (*e.g.* 24-D & 245-T) may cause grotesque new growth with thin twisted leaves and distorted buds. Plants may die in severe cases.

Pre-emergent herbicides contacting the plants' rootsystem via the soil will cause yellowing foliage. Effects of soil borne herbicide may take several years to clear.

NURSERY (HORTICULTURE)

A nursery is a place where plants are propagated and grown to usable size. There are retail nurseries which sell to the general public, wholesale nurseries which sell only to other nurseries and to commercial landscape gardeners, and private nurseries which supply the needs of institutions or private estates. Some retail and wholesale nurseries sell by mail.

Nurseries grow annuals, perennials, and woody plants (trees and shrubs). These have a variety of uses: decorative plants for flower gardening and landscaping, garden vegetable plants, and agricultural plants. Nurseries often grow plants in a greenhouse, a building of glass or in plastic tunnels, designed to protect young plants from harsh weather (especially frost), while allowing access to light and ventilation.

Modern greenhouses allow automated control of temperature, ventilation and light and semi-automated watering and feeding. Some also have fold-back roofs to allow "hardening-off" of plants without the need for manual transfer to outdoor beds.

Some nurseries specialize in one phase of the process: propagation, growing out, or retail sale; or in one type of plant: groundcovers, shade plants, fruit trees, or rock garden plants.

Nurseries remain highly labour-intensive. Although some processes have been mechanised and automated, others have not. It remains highly unlikely that all plants treated in the same way at the same time will arrive at the same condition together, so plant care requires observation, judgement and manual dexterity; selection for sale requires comparison and judgement. A UK nurseryman has estimated that manpower accounts for 70% of his production costs.

Business is highly seasonal, concentrated in spring and autumn. There is no guarantee that there will be demand for the product - this will be affected by temperature, drought, cheaper foreign competition, fashion, etc. A nursery carries these risks and fluctuations. Annuals are sold in trays (undivided containers with multiple plants), flats (trays with built-in cells), peat pots, or plastic pots. Perennials and woody plants are sold either in pots, bare-root or balled and burlaped and in a variety of sizes, from liners to mature trees.

Plants may be propagated by seeds, but often desirable cultivars are propagated asexually by budding, grafting, layering, or other nursery techniques.

CLASSIFICATION OF PLANTS

Plants are classified in several different ways, and the further away from the garden we get, the more the name indicates a plant's relationship to other plants, and tells us about its place in the plant world rather than in the garden. Usually, only the Family, Genus and species are of concern to the gardener, but we sometimes include subspecies, variety or cultivar to identify a particular plant.

Starting from the top, the highest category, plants are classified as follows. Each group has the characteristics of the level above it, but has some distinguishing features.

The further down the scale you go, the more minor the differences become, until you end up with a classification which applies to only one plant.

IMPORTANT FEATURES OF FLOWERING PLANTS

4Flowering plants (Angiospermae) represent one of the largest groups of primary producers. Their contribution to the production of oxygen as well as that to the nutriment of animals and man is consequently very large. All features reviewed in this chapter refer to seed-producing plants, also called spermatophytes.

Typically, flowering plants are organized into an underground root and a shoot above ground that consists itself of a stem and leaves. The organs of a plant that serve sexual reproduction are the flowers. Part of the pollinated flower ripens and becomes the fruit.

In contrast to many other plant groups, flowering plants are striking, numerous and common. They are the most important group of the so-called primary producers that generate the prerequisite for life on earth: oxygen. Green plants have the ability to convert solar energy into chemical energy (photosynthesis) producing the oxygen necessary for all other organisms as a by-product. The useable plants among the flowering plants are - directly or indirectly - the basis of human existence; they are, too, an important economical factor. A basic knowledge of flowering plants should therefore be among everybody's general knowledge.

Much has been written about flowering plants and every reader of this chapter will miss something that he regards worth knowing, while he might find other information trivial. But everybody will understand that it is impossible to review in a few lines a theme about which an extensive, partly popular scientific literature exists. And although this term may sometimes be used in a disparaging way, most of the popular scientific literature is scientifically correct, lucid and, above all, very well illustrated.

To get an idea of the variety of existing plants and to get to know special species, it is necessary to identify them. Many books on classification with different approaches exist. In many popular books, colour photos or drawings are used and often is the colour of the flower a primary feature of recognition. Most of the so-called scientific books on classification work on dichotomic keys, *i.e.* the user is asked a lot of questions in succession and has at each one to decide between two answers. This procedure is continued until the plant has been identified.

The "scientific nature" is mostly based on the completeness of the varieties present in a book, since almost all of the illustrated books contain only the most common or most striking plants. Recently, electronic tools have been used to figure out identification keys for plants. The research into the flora of Central Europe has a tradition that goes back for centuries. It is mirrored in the nearly complete modern floras and books on classification. Incomplete, if existing at all, are books on the classification of less well discovered regions, like tropics, subtropics and many mountain areas.

The question of the origin of the wealth of forms (evolution) is discussed elsewhere where it is also shown that mountains with their rather small and isolated areas provide ideal conditions for the coming into being of new species.

This is the reason, why even very experienced botanists equipped with renowned books on classification of the Central European flora will sometimes and in some places fail (Alps).

This chapter will deal with the characteristics of flowering plants that produce seed (phanerogames or spermatophytes) only. Many of the structures present in this plant group can be found with other non-flowering plants, too. But mosses, ferns and algae miss some features, like flowers or seeds while others, like roots or leaves exist in an incomplete way or are replaced by other organs.

The body of vegetation of many-celled algae (and mosses) is called thallus, that of flowering plants, ferns and fern-like plants (pteridophytes) is called cormus. The latter are therefore summed up as cormophytes. The special features of the different plant groups. The body of vegetation of a "typical" flowering plant consists of an underground root and a shoot above ground. The shoot is organized into stem and leaves. Each of these basic organs can exist in many variations and these again can be combined in many different ways. The almost unlimited ability of combination is one of the main reasons of the existence of such a high number of species while at the same time, the identification of the relations of the species is aggravated.

If seemingly different organs with different functions can be traced back to the same basic organ, they are called homologous. It is also spoken of homology. Contrasting is analogy where organs with a similar look and function have descended from different organs.

THE STEM

Typically, flowering plants are organized into an underground root and a shoot above ground that itself consists of a stem and leaves. The stem is the main supporting axis of a flowering plant. It is made up of nodes, the part where one or more leaves are borne and of the usually longer, in-between lying internodes. Stems are characterized by their type of branching, by their surface and their symmetry. Special cases of stems are those of climbing plants, succulence, runners and rhizomes.

Shoot and root can be clearly distinguished within a seed. After germination, they do develop into the shoot that grows towards the sun and into the root which grows towards the earth's score. The shoot that develops during the first stage of development is called the main- or the primary shoot and depending on the plant, it does bear one, two, or more cotyledons. Growing and sprouted shoots are organized into nodes and the in-between lying internodes.

The nodes are the parts where the leaves are inserted. At the tip of a growing shoot sits the terminal bud of the shoot apex that is surrounded by the leaf (or flower) primordias. In the area of the nodes are axillary buds where the axillary shoots insert.

Vegetable shoots are either annual or perennial. The realised alternative depends usually on the species and in some species on the quality of the environment.

The shoot of annual species is mostly fleshy and is usually called a stalk while perennial plants lignify and thus form trunks and branches. Some species, like Graminaceae or Poaceae have a hollow stalk that is called a blade of grass.

To characterize a shoot, its type of branching and its surface (smooth, hairy, with or without grooves, round or squared) is used, as well as the ratio of heights to diameter.

It can generally be distinguished between monopodial and sympodial branching. Monopodial branching exists, if a continuous main shoot has feeble apical shoots. Sympodial branching is given, if apical shoots of different generations exist at the same time since the apical meristem of each sympodial branch has either died or is differentiated into a flower, a thorn or a tendril and has thus lost its ability to grow. The respective shoot tip is nearer to the ground than that of the next apical shoot. If only one of the apical shoots has the same direction as the main shoot would have had, then a monochasium results which may sometimes be difficult to distinguish from monopodial branching.

But if two buds of a sympodial branch sprout at the same time, a more or less fork-shaped pattern, a dichasium, is formed. Hardly ever do several buds put forth at the same time (pleiochasium).

An extreme ratio of heights to diameter is found in climbing plants where the diameter (and the stability) is very small compared to the length so that upright growing is only possible with the support of other plants or growing aids. Succulence is another extreme, that is typical for plants living in arid or saline habitats.

Most shoots are of a radial symmetry but with a few species, the apical shoots may be changed into flat, leaf-like phylloclades. Often, especially with wooden plants, the axes are differentiated into long and short shoots.

The internodes of the long shoots that cause elongation with trees and shrubs are longer than the short shoots. These again are often specialized apical shoots with shortened internodes. The thorns of white-horn (Crataegus), for example, or those of sloe (Prunus spinosa) correspond to such short shoots. Apical shoots may, too, be changed into runners or rhizomes. Runners are horizontally growing shoots with greatly prolonged internodes that are either underground or above ground and are sometimes bearing leaves. The internodes produce sometimes roots at the part that is nearest to the main shoot (shoot-borne roots). Some parts of these runners can switch to an upright growth and thus behave like a normal shoot.

They serve to propagate vegetatively (strawberries are a typical example). Rhizomes are mostly underground and grow in a horizontal (plagiotrop) manner. They are rather enduring shoots whose internodes usually stay short. They serve as survival organs and are often at the same time used for storage; sometimes, they end in specially developed storage tubers like in potatoes. Except for small, scale-like leaves, rhizomes do usually have no leaves. Shoots which happen to come into touch with a moist soil surface may start to develop leaves and flowers.

ROOTS

Function to take up water, nutriments and minerals from the soil. The primary root is the main axis of a root system. Mostly, it bears thinner side-roots. Special types of primary roots are tap roots or turnips. Adventitious roots are roots originating directly from the stem. The most striking morphological feature of roots is their lack of leaves. Roots are organized into primary roots and side-roots. The latter may develop a frequently branched and extensive root system. Among the prominent types of roots are the tap roots where the primary root is dominant while the side-roots are hardly developed at all. The primary root grows deep into the soil. Swollen roots that have developed into storage organs are called turnips. Their formation may involve the hypocotyl. Swollen side-roots can also become storage organs (tubers).

The root-system of many monocots (like, that of the poaceae,for example) is completely formed from adventitious roots. Adventitious roots are roots that originate directly at the shoot. Rhizomes and shoots that touch the soil are able to form adventive roots. This fact is used to propagate plants vegetatively through cuttings (parts with nodes in them).

Epiphytes do often develop air roots. They are, just like usual roots able to take up minerals and rain water and sometimes they can even perform photosynthesis.

The picture shows a two-dimensional model of a root seeking water in the soil during its development. The initial water distribution has been predetermined, forming an S-shaped zone of high concentration indicated by the light colour. The growing tips of the main root and rootlets absorb water that diffuses in the soil.

The decreased water concentration is indicated by dark areas that emerge around the root system. In areas with insufficient water concentration the rootlets cease to grow before they have reached their potential full length.

A leaf is an aerial and lateral outgrowth of the stem of a usually flat and dorsiventral anatomy. It functions mainly to manufacture food by photosynthesis and consists typically of a stalk also called petiole, a flattened blade, the lamina, and the leaf base. Strands of conducting and strengthening tissues, the veins, run through it. Their pattern, also called leaf venation, is a feature of characterization. Leaves may be simple, *i.e.* undivided or compound (composed of several parts called leaflets). The blade margin and the leaf arrangement at the stem are further features of characterization.

While potentially unlimited growth is a feature of the shoot, that of leaves is usually limited. They are almost always flat and have a dorsiventral anatomy, *i.e.* their upper and their undersurface are constructed differently and have different functions. This morphological and anatomical peculiarity is strongly connected to the main function of leaves, photosynthesis. With most conifers, the leaves are very thin and often tough (needles).

A typical leaf is organized into blade (lamina), stalk or petiole and leaf base. In some plant species, the leaf base is plainly set off against the petiole while the transition is invisible in others. Sometimes, it is merely indicated by a faint widening. Occasionally, lateral excrescences develop. Most species have clearly recognizable petioles. If they are missing, the leaves are said to be sessile in which case the leaf base may sometimes surround the shoot.The extremes are laminas with fused basal rims evoking the impression of a leave that the shoot grows through (Claytonia perfoliata). The basis of the leaf stalk may even grow down on the shoot a little bit further. With many monocots, like Poaceae or Graminaceae, the shoot is completely surrounded by a sheath-forming petiole.

The blade is radiated by leaf veins that contain the fibres. It is distinguished between parallel, arched, pinnated or netlike vein structures.

Just as variable as the vein structure is the shape of the leaves. The fundamental difference is that between simple and compound leaves. The compound leaves are built from several small leaves or from pinnations that sit in a regular organisation at the undivided or branched rhachis.

Cataphylls are simply organized, scale-like leaves. They can be found at the sprout, at the basis of sprouting shoots or underneath the normal leaves and in wooden plants also as a protection for the buds.

Finally, bracts have to be mentioned that live above the region of the usual leaves, for example at the inflorescence as subtending bracts for the flowers. But they can also have very different functions.

Remarkable are the strikingly coloured bracts of Bougainvillea spectabilis and Poinsettia pulcherrima (poinsettia) that add strongly to the attractivity of the rather inconspicuous flowers.

Another example are the Cuckoo-pint and its relatives where the bract encloses the inflorescence and is at the same time a refined trap for pollinating insects (Arum maculatum).

Some species, especially those that live in water have differently shaped leaves (heterophylly). An example is Ranunculus aquatilis (water crowfoot) which has pinnation-shaped, submerse living leaves and flat swimming leaves.

Another feature of recognition is the blade margin that can be shaped very differently. It can be plain, serrated, double serrated, crenated or sinuated. Leaves with deep clefts are called pinnatified, pectinated, palmated or lobed.

Quite a few species have structures called leaf metamorphosis (variations),because they are homologous to leaves. Among these are thorns where themoulding of the blade does not occur and the leaf veins have been strengthened by the deposit of a though material, as well as leaf tendrils that also miss the lamina but have an extremely pliable and tension-proof petiole. There may, too, be succulence of leaves.

Leaves are always organized around the shoot in a regular pattern that can be recognized best when looked at from above. It is obvious in most cases that the leaves are arranged in the way of a screw and that there is always the same angle measurable between subsequent leaves (angle of divergence).

PROGRAMME IMPLEMENTATION FOR 2005-06

- Thrust on Hi-Tech Horticulture and precision farming.
- Expansion of area under micro-irrigation and fertigation.
- Stabilizing the area of water loving crops and expanding the area under dry land crops with a focus on effective water management.
- Area expansion of horticulture crops through various Waste Land Development programmes.
- Strengthening the system of production of pedigree and hybrid planting materials in both public and private sectors.
- Promotion of Organic farming with focus on export market.
- Promotion of Agri Export Zones (AEZ) for specified crops.
- Building up of Public and Private Partnership.
- Promotion of Contract / Corporate farming.
- Empowerment of farmers with special focus on farm women.
- Effective transfer of technologies by tour-cum-training to farmers.
- E-Governance and Human Resources Development through effective training for extension officers.
- Linkage with Agro Processing Industries with New Anna Marumalarchi Thittam (NAMT).
- Post harvest management and reduction of post harvest losses.

APPLICATIONS OF BIOTECHNOLOGY IN HORTICULTURE

The requirement of fruits and vegetables is increasing proportionally with the increasing population in the country.How do we keep horticulturalproduction on par with the burgeoning population? Although conventional plant breeding techniques have made considerable progress in the development of improved varieties, they have not been able to keep pace with the increasing demand for vegetables and fruitsin the developing countries.

Therefore an immediate need is felt to integrate biotechnology to speed up the crop improvement programmes.Biotechnological tools have revolutionized the entire crop improvement programmes by providing new strains of plants, supplyof planting material, more efficient and selective pesticides and improved fertilizers. Many genetically modified fruits and vegetables are already in the market in developed countries.Modern biotechnology encompasses broad areas of biology from utilization of living organisms or substances from those organisms to make or to modify a product, to improve plant or animal or to develop micro-organisms for specific use.It is a new aspect of biological and agricultural science which provides new tools and strategies in the struggle against world's food production problem. The major areas of biotechnology which can be adopted for improvement of horticultural crops are –

TISSUE CULTURE

One of the widest applications of biotechnology has been in the area of tissue culture and micro propagation in particular. It is one of the most widely used techniques for rapid asexual in vitro propagation. This technique is economical in time and space affords greater output and provides disease free and elite propagules.

It also facilitates safer and quarantined movements of germplasm across nations.When the traditional methods are unable to meet the demand for propagation material this technique can produce millions of uniformly flowering and yielding plants.Micropropagation of almost all the fruit crops and vegetables is possible now.Production of virus free planting material using meristem culture has been made possible in many horticultural crops.Embryo rescue is another area where plant breeders are able to rescue their crosses which would otherwise abort.

Culture of excised embryos of suitable stages of development can circumvent problems encountered in post zygotic incompatibility.This technique is highly significant in intractable and long duration horticultural species.Many of the dry land legume species have been successfully regenerated from cotyledons, hypocotyls, leaf, ovary, protoplast, petiole root, anthers, etc.,Haploid generation through anther/pollen culture is recognized as another important area in crop improvement.It is useful in being rapid and economically feasible.Complete homozygosity of the offspring helps in phenotype selection for quantitative characters and particularly for qualitatively inherited characters making breeding much easier successful isolation, culture and fusion of plant protoplasts has been very useful in transferring cytoplasmic male sterility for obtaining hybrid vigour through mitochondrial recombination and for genetic transformation in plants.

In vitro germplasm conservation is of great significance in providing solutions and alternative approaches to overcoming constrains in management of genetic resources.In crops which are propagated vegetatively and which produce recalcitrant seeds and perennial crops which are highly heterozygous seed storage is not suitable.In such crops especially, *in vitro* storage is of great practical importance.These techniques have successfully been demonstrated in a number of horticultural crops and there are now various germplasm collection centers.In vitro germplasm also assures the exchange of pest and disease free material and helps in better quarantine.

Plant breeders are continually searching for new genetic variability that is potentially useful in cultivar improvement. A portion of plants regenerated by tissue culture often exhibits phenotypic variation atypical of the original phenotype.Such variation, termed somaclonal variation may be heritable i.e. genetically stable and passed on to the next generation.Alternatively, the variation may be epigenetic and disappear following sexual reproduction.These heritable variation are potentially useful to plant breeders.Genetic Engineering of Plants

Using techniques of genetic engineering many useful genes have been introduced into plants and many transgenic plants have been developed in which the foreign DNA has been stably integrated and resulted in the synthesis of appropriate gene product. Transgenic plants have covered about 52.6 m hectares in the Industrial and developing countriesupto 2001. Genes for the following traits have been introduced to the crop plants.

Herbicide tolerance: Transgenic plants are developed that are resistant to herbicides allowing farmers to spray crops so as to kill only weeds but not their crops.Many herbicide tolerant plants have been developed in tomato, tobacco, potato, soybean, cotton, corn oilseed rape,petunia, etc. Glyphosate is one of the most potent broad spectrum environment friendly herbicide known, it is marketed under the trade name Round up.Glyphosate kills plants by blocking the action of an enzyme (5-enolpyruvyl shikimate-3-phosphate synthase) (EPSPS) an essential enzyme in the biosynthesis of aromatic amino acids, tyrosine, phenylalanine and tryptophan.Amino acids are building blocks of protein.

Transgenic plants resistant to Glyphosate have beendeveloped by transferring gene of EPSPS that over prodoce this enzyme thus inhibiting the effect of Glyphosate.A number of detoxifying enzymes have been identified in plants as well as in microbes.Some of these include glutahthione-s-transferase or GST in maize and other plants which detoxifies the herbicidebromoxynil and phosphinothricin acetyl transferase (PAT) which detoxifies the herbiside PPT (L-phosphinothricine).Transgenic plants using *bxn* gene from Klebsiella and *bar* gene from Strepotomyces have been obtained in potato, oilseed, sugarbeet, soybean, cotton and corn and are found to be herbicide resistance.These transgenic plants reduce the use of weeding labour,farmers costand increase yield.

Engineeringpathogen resistance: Viruses are the major pests of crop plants which cause considerable yield losses.Many strategies have been applied to control virus infection using coat protein and satellite RNA.Viruses are submicroscopic pockets of nucleic acid (DNA or RNA) enclosed in a protein coat and can multiply within a host cell. Use of viral coat protein as a transgene for producing virus resistant plants is one of the most spectacular successes achieved in plant biotechnology.Coat protein gene from tobacco mosaic virus (TMV) classified as a positive strand RNA virus has been transferred to tobacco, making it nearly resistant against TMV.Using gene for nucelocapsid protein resistance has been introduced in crops like tomato, tobacco, lettuce, groundnut, pepper and in ornaments like *Impatiens*, *Ageratum* and *Crysnathemum* againsttomato spotted wilt virus. Use of satellite RNA (SATRNA) makes many transgenic plants resistant toCucumber Mosaic Virus (CMV).Transgenic resistant plants have also been developed against alfalfa mosaic virus, potato virus X, Rice tungro virus, tobacco rattle virus and Papaya ring spot virus.

During the last decade many resistance genes whose products are involved in recognizing the invading pathogens have been identified and cloned.A number of signaling pathways which follow the pathogen infection have been dissected.Many of the antifungal compounds synthesized by plants which combat fungal infections have been identified.The major strategies for developing fungal resistance have been production of transgenic plants with antifungal molecules like proteins and toxins,and generation of hypersensitive response through R genes or by manipulating genes of SAR pathway.A chitinasegene from bean plants in tobacco and *Brassica napus* showed enhanced resistance to *Rhizoctonia solani*. In another case chitinase gene obtained from *Serratia marcescens*(soil bacterium) is introduced in tobacco making it resistant to *Alternaria longipes* which causes brown spot diseases.Acetyl transferase gene is introduced in tobacco making it resistant to*Pseudomonas syringea*, a causal agent of wild fire disease.

Stress resistance : A number of genes responsible for providing resistance against stresses such as to water stress heat, cold, salt, heavy metals and phytohormones have been identified.Studies are also being conducted on metabolites like proteins and betains that have been implicated in stress tolerance. Resistance against chilling was introduced into tobacco plants by introducing gene for glycerol-1-phosphate acyl-transferase enzyme from *Arabidopsis*.Many plants respond to drought stress by synthesizing a group of sugar derivatives called polyols (Mannitol, Sorbitol and Sion) .Plants that have more polyols are more resistant to stress.Using a bacterial genecapable of synthesizing mannitols it is possible to raise the level of mannitol very highmaking plants resistant to drought.

Fruit Quality: Tomatoes which ripen slowly are helpful in transportation process.Transgenic tomato with reduced pectin methyl esterase activity and increased level of soluble solids and higher pH increases processing quality.Tomatoes exhibiting delayed ripening have been produced either by using antisense RNA against enzymes involved inethylene production (Eg ACC synthase) or by using gene for deaminase which degraded l-aminocyclopropane-l-carboxylic acid (ACC) an immediate precursor of ethylene.This increases the shelflife of tomatoes. These tomatoes can also stay on the plant long giving more time for accumulation of sugars and acids for improving flavour.It is produced at commercial level in European and American countries.Tomatoes with elevated sucrose and reduced starch could also be produced using sucrose phosphate synthase gene.Starch content in potatoes has been increased by 20-40% by using a bacterial ADP glucose pyrophosphorylase gene.

Pest resistance: The insecticidal beta endotoxin gene (bt gene) has been isolated from *Bacillus thuringiensis* the commonly occurring soil bacteria and transferred to number of plants like cotton, tobacco, tomato, soybean, potato, etc. to make them resistant to attack by insects.These genes produce insecticidal crystal proteins which affect a range of lepidopteran, coleopteran , dipteran insects.These crystals upon ingestion by the insect larvaare solubilised in the highly alkaline midgut into individual protoxins which vary from 133 to 136 kDa in molecular weight.Insecticidal crystal proteinproduced during vegetative growth of the cells (VIP)arealso found to be highly effectiveagainst insect control.Bt resistant plants are already in the market.

Male sterility and Fertility restoration: This is helpful in hybrid seed production.Transgenic plants with male sterility and fertility restoration genes have become available in *Brassica napus*.It facilitates production of hybrid seed without manual emasculation and controlled pollination as often done in maize.In 1990, Mariani and others from Belgium have successfully used a gene construct having another specific promoter from TA29 gene of tobacco and bacterial coding sequence for a ribonuclease gene from Bacillus Sp. (barnase gene) for production of transgenic plants in *Brassica napus*.Here the translated gene prevented normal pollen development leading to male sterilily.

MOLECULAR DIAGNOSTICS

Nucleic acid probes:- It is now possible to detect the plant diseases even before onset of symptoms by using cDNA probes.Probes are nucleic acid sequences of pathogen causing organisms labeled with certain markers.cDNA probes corresponding to specific regions of the pathogens can be generated using standard recombinant DNA technique.

Monoclonal antibodies (McAb): Immunochemical techniques are extremely useful for the rapid and accurate routine detection of plant pathogens and ultimately the diagnosis of plant disease and their relatedness,The introduction of hybridoma technology has provided methods for the production of homologous and biochemically defined immunological reagents of identical specificity which are produced by a single cell line and are directed against a unique epitope of the immunizing antigen. The great potential of McAbs in phytopathological diagnostics is essential because of homogeneous antibody preparations with defined activity and specificity can be produced in large quantities over long periods.Even though hybridoma technology is a laborious and expensive enterprise compared to standard immunization procedures it is going to be widely used for large scale diagnosis.

MOLECULAR MARKERS

The possibilities of using gene tags of molecular makers for selecting agronomic traits has made the job of breeder easier.It has been possible to score the plants for different traits or disease resistance at the seedling stage itself.The use of RFLP (Restriction Fragment Length polymorphism), RAPD (Random Amplified Polymorphic DNA) , AFLP (Amplified Fragment Length Polymorphism) and isozyme markers in plant breeding are numerous. RFLPs are advantageous over morphological and isozyme markers primarily because their number is limited only by genome size and they are not environmentally or developmentally influenced.Molecular maps now exist for a number of crop plants including corn, tomato, potato, rice, lettuce, wheat, Brassica species and barley.RFLPs have wide ranging applications including cultivar finger printing, identification of quantitative trait loci, analysis of genome organization, germplasm introgression and map-based cloning.AFLPis becoming the tool of choice forfingerprinting because of its reproducibility compared to RAPD.Microsatellile or simple sequence repeats (SSRS) markers have also become the choice for a wide range of applications in genotyping, genome mapping and genome analysis.

DEVELOPMENT OF MICROBIAL INOCULAN

Indiscriminate and injudicious use of chemical fertilizers and pesticides for the crop production and control of insect-pests has resulted in pollution of the environment deterioration of soil health and development of resistance by many insects and residue problems.Hence there is a great concern world wide to use safer biofertilisers and biopesticdies in the integrated nutrient management and pest management systems.

Biofertilizers are micro-organisms which fix atmospheric nitrogen or solubilise fixed phosphorus in the soil and make more nutrients available to the plant.Some of the organisms providing major inputs are the biological nitrogen fixing organisms like *Rhizobium, Azotobacter, Azospirillum* and phosphate solubilising organisms like *Bacillus polymyxa, B. magaterium, Pseudomonas striata* and certain fungal species of Aspergillus and *Penicillium.*

The benefits of using micro-organisms as fertilizers are many fold. They are less expensive, nontoxic to plants, do not pollute the ground water nor render the soil acidic and unfit for growth of plants. *Rhizobium* forms nodules on the roots of leguminous plants and help in fixing nitrogen from the atmosphere to ammonium irons which get converted to amino acids in the plant system.Inoculation with this bacteria helps in reducing addition of nitrogenous fertilizers to the soil.*Azospirillum* is also found colonizing inter cellular spaces inside the root system.These bacteria also contribute substantially to the nitrogen requirement of the plant.

Phosphate solubilising bacteria are another group of micro-organisms which solubilise the insoluble phosphorus in the soil and make them readily available to the crop.

Mycorrhiza is the symbiotic association of the roots of crop plants with non-pathogenic fungus.They provide nutrients absorbed from deeper layers of soil to the plants.They help the plants in better plant establishment and growth when inoculated.Many fruit crops like papaya, mango, banana, citrus, pomegranate are found to be dependent on this association and are greatly benefited by its inoculation in procuring higher phosphate and other nutrient from the soil.These mycorrhizal associations help the plants in overcoming pathogen attack also.They improve soil characters too.

Genetic modification of microbes: By using DNA recombination technique it has been possible to genetically manipulate different strains of these bacteria suitable to different environmental conditions and to develop strains with traits with capacity for better competitiveness and nodulation. Biopesticides are biological organisms which can be formulated as that of the pesticides for the control of pests.

Biopesticides are gaining importance in agriculture, horticulture and in public heatlh programmes for the control of pests.The advantages of using biopesticides are many.They are specific to target pests and do not harm the neo- target organisms such as bees, butterflies and are safe to humans and live stocks, they do not disturb the food-chain nor leave behind toxic residues. Some of the microbial pesticides used to control insect pests are *Bacillus thuringiensis* species to control various insect pests.Insecticidal property of these bacteria are due to crystals of insecticidal proteins produced during sporulation.These proteins are stomach poisons and are highly insect specific.Bt toxins could kill plant parasitic nematode too.Number of baculoviruses (BV) nuclear polyhedrosis virus (NPV) is being developed as microbial pesticides both nationally and internationally, A few examples of these are *Heliothis, Spodoptera, Plusia, Agrotis, Trichoplusia*, etc.

Biocontrol agents : These are other microbes which are antagonistic to several pathogenic fungus and are good substitutes to fungicides or insecticide.These are Bacillus sps. *Pseudomonas fluorescens, Trichoderma, Verticillium sp., Streptromyces* sps. etc. These organisms are commercially available.

8

Inflation in Agriculture Sector

THE ESTABLISHMENT OF FARMING

Except in nations with Communist governments, most farm land is privately owned. This does not mean, however, that the land is owned by those who farm it. In most countries a major aspiration of farm people has been to achieve the ownership of the land they work.

After World War II, Japan and Taiwan underwent land reforms that were intended to broaden ownership; these are generally considered to have been highly successful. Similar reforms have been advocated in other countries. On a cooperative farm the land is owned jointly by the members of the group who farm it.

The cooperative generally also owns all the major means of production, and the members supply all or most of the labour. While there are examples of cooperative farms in many countries, they loom large only in Israel, where the kibbutzim control about a fifth of the total agricultural land.

In a collective farm, at least as organized in the former Soviet republics, the land was owned by the state but was permanently leased to the *kolkhoz* (collective farm).

The *kolkhoz* owned its own equipment and livestock and was required to meet certain commitments to the state in the form of deliveries of farm products. In theory the members of the *kolkhoz* were to elect the officers of the farm and establish the procedures by which the net product was to be divided among the members for services performed.

In practice, however, their autonomy was severely limited by the economic plans. In most cases these plans were incredibly detailed, specifying the crops to be grown, the times of plowing, planting, and harvesting, the quantities of fertilizer and manures to be used, and the kinds of livestock to be maintained.

On state farms the land and all other means of production are owned by the state. The workers are paid in wages, and management decisions are made by individuals directly responsible to the state.

KINDS OF FARM OPERATION

If a family farm is defined as one for which the farm operator and members of his family supply at least half of the labour, the majority of farms in the world are family farms. Family farming is carried on under a wide range of conditions, from the small farms of Asia to the highly mechanized farms of Canada, the United States, and the United Kingdom.

The family farm may be owned by the farmer or rented. The most rapidly expanding type of tenure in the United States is that in which the farmer owns part of the land and rents the remainder; almost one-third of all farmland in the United States consists of part-owner farms. This arrangement enables the farmer to increase the size of the farm through renting and to invest capital in machinery and livestock.

Family farms may be large in terms of total assets or sales. The relative importance of family farms among the largest farms in the United States has increased over the past few decades. One of the more striking changes in industrial countries has been the increased importance of nonfarm income received by farm families.

In the United States, Canada, and Japan more than half of the total income of farm families comes from nonfarm sources, while in most western European countries at least a third of the income of farm families is earned outside of agriculture.

A system of tenant farming known as sharecropping developed in the South of the United States following the freeing of the slaves in the 19th century.

It was essentially an adjustment of the plantation system created to permit the owners to maintain a large measure of control over farm operations. The sharecropper usually supplied only the labour, while the owner provided animal power, machinery, and most of the other inputs in the form of an advance.

The sharecropper received what was left after he had paid back the owner—generally about half of what had been produced.

For various reasons, including the exodus of blacks from American agriculture, the introduction offarm machinery, and the reduction in the acreage of cotton, the number of sharecroppers in the South has diminished by well over 80 percent since 1935.

In the past several decades there has been a growth of large-scale farming run as a business enterprise. These "industrial farms" are of growing significance in world agriculture.

There are farms covering extensive areas of land in Africa, South America, and Australia, but most of them do not rely heavily upon machinery or other purchased inputs.

Farms in the United States are becoming larger as their numbers grow smaller. Such large farms tend to specialize in the production of vegetables, fruits, cotton, poultry and poultry products, and livestock.

Comparative Strengths and Weaknesses

If they were free to choose, most farm families would want to own the land they farm. Wherever collectivization of private farmers has been carried out, it has required the use of force or the threat of force.

But if family farming is to be viable, it must function efficiently, which means that farmers must have access to adequate sources of credit; must be able to obtain fertilizers, machinery, and other equipment; and must be able to market their produce easily. Laws and institutions must be sufficiently flexible to permit the average size of farms to increase as economic growth occurs.

Collective farming did not fulfil the hopes of its early advocates. In the Soviet Union the collective farm was used by Stalin as a means of exploiting the rural population in order to finance the expansion of industrialization.

In the post-Stalin era the incomes of collective farm members increased, and it was believed that many remaining difficulties could be eliminated if the farms were given greater freedom in running their affairs.

Nothing in the concept of the collective farm required the imposition of delivery quotas, centralized control of farm investment, or a particular organization of farm labour. Another weakness of collective farms was the failure to provide adequate incentives for individual members. Because of the difficulties involved in rewarding members for their individual work on the common land, the household plots of the members all too often tended to flourish at the expense of the collective.

There is no ideal form of organization that fits all farming. Under some circumstances the ownership of land may absorb so much capital that other investments, such as machinery and livestock, are neglected. Land rental may be a better alternative for many families, especially those with limited capital. The Israeli kibbutz has made it possible for many people with little or no agricultural experience to learn farming techniques quickly and efficiently. The most important thought is whether the other institutions—economic, political, and social—are adequate to provide farmers with a wide range of resources and alternatives.

INFLATION IMPACT ON AGRICULTURAL SECTOR

Inflation is regarded as one of the greatest problems that affect any international economy, as it reflects extra economical and social burdens, besides its obstructions that lead to a misleading and negative economy.

So that, this study aimed to measure the current inflation rates in Saudi Arabia during the periods of 1970 up to 2003, and its effects on growth rates including the agriculture sector and its related investments, also on the supply and demand of the main agricultural commodities in Saudi Arabia. The study also noted and realized the importance of measuring the inflation rates, and its effect on the Saudi economy, cost living index, and wholesale price index and implicit index including the monetary stability factor; excess demand, net excess demand and exceeding monetary.

This study also realized the importance of analysing the inclinations during the process of measuring the inflation effect on supply, demand of the most important agriculture commodities and its effect on the economical growth and agricultural investments.

As the study relied on analysing the time series data, as it used Dubki Foller advanced test (ADF) in determining stability of variables, besides estimating the cointegration equations used in two steps Angle Granger method and Johansen's method. Also, this study has relied on some published data in different annual reports issued by the Saudi Arabia Monetary agency about the achievements of development plans, the annual statistical.

IRRIGATION WATER PRICING AND FARM PROFITABILITY IN THE BURDEKIN DELTA SUGARCANE AREA

Water is becoming an increasingly scarce resource and therefore constraining agricultural development in many regions and countries of the world. In the past, new physical systems have been built to harness additional water resources when demand for water exceeded supply.

However, with increasing demand by users in agriculture as well as from other industries, and increasing concern within the general community about potentially harmful externalities of irrigation, emphasis is now being placed on improving the performance of existing irrigation systems. In Australia, efficient and sustainable management of water resources is increasingly becoming a policy objective (Thomas, 1999).

There are several factors to be considered in irrigation management to improve productivity and reduce costs. One of the key decisions is how much water should be allocated to a particular crop relative to other crops.

This decision needs to be based on the quality and availability of water resources, reliability of water supply, physiological requirements of the crop, and the expected value of crop output.

For a canegrower, many possible strategies can be followed when deciding the optimal level of irrigation water to apply. A frequently followed strategy is to apply water to an individual block of cane at a level that achieves maximum crop yield.

This approach may be used when there is no restriction on irrigation water supplies and little extra cost involved in the application of additional water. However, when water supplies are limited, or when additional marginal quantities of water incur a rising cost, it is obviously better to provide the optimal or efficient level of irrigation water across various crops on the farm, in order to achieve a higher overall level of farm profitability. The efficient allocation of water must strike a balance between competing crops on the farm and crops on different soil types. This requires an economic modelling approach which can determine the optimal allocation of water for various crops and soil types.

This chapter briefly describes an optimisation model used to estimate private (farmer) profitability of irrigated sugarcane on farms with multiple soil types. This model is designed to aid farm managers in decision making to achieve optimal irrigation levels for farms with various soil types in one of the major canegrowing areas in Australia.

The paper then examines the impact on farm profitability and optimal level of irrigation due to changes in water charges, pricing structures and sugar price for a representative sugarcane farms in both the North and South Burdekin Water Board Areas of the Burdekin delta. In the following section, the theoretical concepts underpinning the determination of economically optimal irrigation rates are discussed. Issues pertaining to the case study area and need for effective irrigation management are then presented.

ECONOMICALLY OPTIMAL WATER USE: THEORETICAL BACKGROUND

Economic efficiency refers to the combination of inputs that achieves individual or social objectives to the greatest possible extent in relation to individual farms and to society. In the case of a single variable input, economic efficiency relies on the location of the production function on which the farm is operating. The most profitable point of operation is obtained either by determining the most profitable amount of input or the most profitable level of output. Either method results ultimately in the same optimal level of the variable input because the production function relates input to output in a unique manner. The maximum profit does not occur where physical output is at a maximum if there is a price attached to the input.

This is because a stage of diminishing marginal product from a single variable input (such as irrigation water) usually exists and, beyond the point of maximum profit, the added units of input cost more than they are able to earn. Thus the quest for maximum yields is not always an economically sensible goal, and at times can also be less than optimal environmentally. Using inputs beyond the point of greatest efficiency is not consistent with profit maximisation unless all inputs into the production process are free (Doll and Orazem, 1984). This principle applies when examining both private and social efficiency. Farm production processes are considered efficient when production inputs are allocated optimally.

However, if the expenditure on an input (irrigation water) is low compared to other costs of production, then the optimal quantity will approximate that leading to maximum crop yield. This situation is particularly likely to occur when inputs are provided at less than their full cost, possibly because they are underpriced or because an important cost element, such as the impact on the environment, has been overlooked. Optimal use of an input occurs when the marginal value product (MVP) equals the cost per unit of the input (P_X). If MVP is lower than P_X, the resource is over-utilised and reducing the quantity used at the current price will increase the MVP towards optimality. On the other hand, if MVP is greater than P_X, the resource is over-utilised and using more of it will not bring additional gains to the grower.

For farmer decision making, the price of the input P_X includes only private costs, not social costs. In the past, many irrigation schemes operated on the basis of recovering operating costs only.

Increasingly, prices for irrigation water are being increased to include two additional components: charges for the use of the capital works involved in the delivery of irrigation water and some payment to the community for the use of the environmental resource. In other words, some of the externalities involved in the use of irrigation water are being borne by irrigators. Thus farmers who use water delivered by public schemes are being asked to pay more for the input as part of a strategy to achieve greater efficiency and more equity in the use of resources.

In situations where there are a number of crops, efficiency dictates that the marginal net benefit (marginal revenue of product less marginal cost of input) be equalized for all crops.

If marginal net benefits are not equal, it is always possible to increase aggregate benefits by transferring water from those crops with low marginal net benefits to those with higher marginal net benefits provided there is enough flexibility to allow water reallocation. The benefits given up by those crops losing water are less than those gained through crops receiving additional water. This procedure results in allocative efficiency (Samuelson and Nordhaus, 1987). At this stage, water is efficiently allocated and no other outcome can improve the welfare of the water user.

If a farm or region has various soil types, then similar economic principles apply in the allocation of water to each soil type, and maximisation of farm profit is achieved only if there is an optimal level of irrigation on each soil type and marginal productivity of water use is the same across all soil types.

Efficiency dictates that the marginal net benefit be equalized for all sites or soil types for each crop. When there is a single crop dominant in a region (due to either economic or agronomic conditions), the optimal level of water application is still achieved when the cost of one additional unit of water is equal to the additional revenue it generates.

In situations such as the Burdekin delta where only one crop, sugarcane, is predominantly grown, there may currently be no competition for the irrigation water from other crops or industries, but the threat of competition from new non-agricultural users has to be considered. If water is an underpriced resource in the region, it will ultimately attract new users (other crops or non-agricultural activities depending on water productivity) into the area and it is better for the farmers concerned to be making the necessary adjustments well ahead of that eventuality.

Mathematical Models in Irrigation Planning and Water Pricing Policies

Mathematical programming models can be used to determine optimal activity and resource input levels.

These models are quantitative in the sense that they indicate, for example, the number of units of each commodity or output to be produced and number of units of each input to be used if a stated goal such as profit maximisation is to be attained. Linear programming (LP) has been used extensively in the analysis of water resource systems and irrigation planning. Some examples are now presented.

Kheper and Chaturvedi (1982) applied an LP model to make decisions about groundwater management options in conjunction with optimal cropping pattern and water production functions. Panda and Kheper (1985) used deterministic LP and chance-constrained LP to maximise net return from irrigation planning.

Paudyal and Gupta (1990) solved the complex problem of irrigation management in a large heterogeneous river basin by using a multilevel optimisation technique.

They determined the optimal cropping patterns in various sub-areas of the basin, the optimal design capacities of irrigation facilities, including surface and groundwater resources, and the optimal water allocation policies for conjunctive use. Mainuddin *et al.* (1997) used an LP model to determine the cropping pattern to ensure optimal utilisation of available land and water resources in a groundwater irrigation project.

Mathematical programming models have also been used to determine demand for water and water pricing policies and impact of these policies on water demand. Chewings and Pascoe (1988) used LP to simulate water use in the Murray Valley in New South Wales, Australia.

They investigated the major consequences of various water pricing policies, subject to resource availability, costs and commodity prices as well as deriving demand functions for water in the valley, by parametising water price changes over a specified range. Tisdell (1996) used an LP model to estimate demand functions for water in Border Rivers region of Queensland by varying water charges.

Ulibarri *et al.* (1998) estimated the change in farm profitability from imposing full-cost federal water and power rates using an LP model in California. Varela-Ortega *et al.* (1998) used a multi-period mathematical programming model to analyse the effect of various water pricing policies on water demand, farmers' incomes and the revenue collected by a government agency in Spain. Amir and Fisher (1999) used an optimising model to analyse production under various water quantities, qualities and pricing policies, and determined water demand curves for various districts in Israel.

Berbel and Gomez-Limon (2000) applied an LP model to three farms in different irrigation units and examined the impact of water price policy on farm income and on regional employment in Spain. They also determined water demand curves that reflected farmers' adaptation to rising costs of production inputs.

Case Study: Burdekin Delta Irrigation Area

The Burdekin River delta is located approximately 90 km south of Townsville on the northeast coast of Australia, close to wetlands, waterways, estuaries and the Great Barrier Reef. It covers an area of about 850 km and together with the Haughton River – Barratta Creek system, is one of the largest alluvial aquifer systems in Australia.

The area has a tropical climate and seasonal rainfall (two thirds of the annual rainfall ranging between 250 mm and 2500 mm) with an average of about 1000 mm, with most falling during January to March.

Evaporation rates vary between 10 mm/day in November and 2.8 mm/day in June. The delta is predominantly used for sugarcane production, with some smaller areas under tropical fruits and vegetables.

There are small areas where groundwater or soil quality is not suitable for the sugarcane crop or horticulture, and these are used for cattle grazing. The delta is one of the few areas in Queensland where cane is grown under full irrigation and conjunctive use of groundwater and surface water is common practice.

The North and South Burdekin Water Boards (NBWB and SBWB) were established in 1965 and 1966 respectively, to replenish and manage the underground aquifers. Artificial recharge is still practised in the delta, although the Water Boards (particularly the NBWB) are promoting a practice of more efficient water use requiring less artificial recharge.

Sugarcane is the dominant crop in the delta with 21,800 ha planted in the NBWB area and 13,344 ha of cane in the SBWB area in 2001, and with only 1,500 and 84 hectares, respectively, under other crops (NBWB, 1998; SBWB, 1999).

One way to achieve higher efficiency in water use is to ensure that the amount applied to cane is matched to soil types and crop demand.

More water-efficient practices will reduce pumping costs for growers and increase farm profitability as well as reducing the potential risk of leaching nutrients and pesticides into the aquifer.

Therefore, it is important to design and implement practices which ensure the long-term viability of irrigated agriculture in the area.

Analytical Framework

The integrated approach applied in this analysis includes using a biophysical simulation model to predict crop yields of sugarcane, linked to a linear programming model to determine the levels of irrigation for various soil types which maximise farm profitability for a representative cane farm in the study area.

Biophysical Models to Determine Crop Yield

For a comprehensive economic analysis, biophysical information such as crop yield is necessary. The APSIM cropping systems model (McCown *et al.* 1996) has been used to obtain these data by linking a sugar crop module, a soil water module, a soil nitrogen module and a surface residue module, as described by Probert *et al.* (1997).

The model 'APSIM-Sugarcane' was configured to simulate continuous cropping over a 20-year period from 1975 to 1995 with a cycle consisting of one plant crop followed by three ratoon crops. A series of crop simulations have been performed to estimate response to applied irrigation water for a range of soil types, water allocation levels and application efficiencies, under the widely used furrow irrigation system. The water inputs and yields obtained from the 20-year simulation runs have been averaged. A selection of irrigation options were chosen for the investigation based on combinations of soil types, water quantities (0 to 35 ML/ha in 1 ML/ha increments) and above-ground application efficiencies of 30% to 60% for three soil types, namely clay, silt and sand.

The three soil types were selected to represent profiles with sharply contrasting levels of plant-extractable soil water (representative of low, medium and highly permeable soil types in the study area).

The long-term climate files for the study area from 1975 to 1995, consisting of daily rainfall, minimum and maximum temperature and solar radiation data, were used as inputs for the simulation analyses.

These files are comprised of a combination of recorded weather station data and data from generated historical meteorological surfaces, obtained from the Bureau of Meteorology and the Queensland Centre for Climate Applications.

Simulated yields were about 20% higher than the average sugar yield in the region because the model does not take into account losses associated with pests, diseases, weed competition and impact of waterlogging or unusual climatic events, and yields are based on uniform soil characteristics.

Therefore, the yields for the various irrigation levels have been reduced by 20% for each of the three soil types. Average water inputs and yields simulated over the 20 years have been used in the economic analysis.

EFFECT ON FOOD PRICES

Some environmentalists, such as George Monbiot, have expressed fears that the marketplace will convert crops to fuel for the rich, while the poor starve and biofuels cause environmental problems.

Environmental groups have raised concerns about this trade-off for several years. The food vs fuel debate reached a global scale in 2008 as a result of the international community's concerns regarding the steep increase in food prices. On April 2008, Jean Ziegler, back then United Nations Special Rapporteur on the Right to Food, called biofuels a *"crime against humanity"*, a claim he had previously made in October 2007, when he called for a 5-year ban for the conversion of land for the production of biofuels. Also on April 2008, the World Bank's President, Robert Zoellick, stated that *"While many worry about filling their gas tanks, many others around the world are struggling to fill their stomachs. And it's getting more and more difficult every day."*

Luiz Inácio Lula da Silva gave a strong rebuttal, calling these claims *"fallacies resulting from commercial interests"*, and putting the blame instead on U.S. and European agricultural subsidies, and a problem restricted to U.S. ethanol produced from maize. The Brazilian President has also claimed on several occasions that his country's sugar cane based ethanol industry has not contributed to the food price crises.

A report released by Oxfam in June 2008 criticized biofuel policies of rich countries as neither a solution to the climate crisis nor the oil crisis, while contributing to the food price crisis.

The report concluded that from all biofuels available in the market, Brazilian sugarcane ethanol is "far from perfect" but it is the most favourable biofuel in the world in term of cost and greenhouse gas balance. The report discusses some existing problems and potential risks, and asks the Brazilian government for caution to avoid jeopardizing its environmental and social sustainability. The report also says that: *"Rich countries spent up to $15 billion last year supporting biofuels while blocking cheaper Brazilian ethanol, which is far less damaging for global food security."*

A World Bank research report published on July 2008 found that from June 2002 to June 2008 *"biofuels and the related consequences of low grain stocks, large land use shifts, speculative activity and export bans"* accounted for 70-75% of total price rises. The study found that higher oil prices and a weak dollar explain 25-30% of total price rise.

The study said that *"...large increases in biofuels production in the United States and Europe are the main reason behind the steep rise in global food prices"* and also stated that *"Brazil's sugar-based ethanol did not push food prices appreciably higher"*. The report argues that increased production of biofuels in these developed regions were supported by subsidies and tariffs on imports, and considers that without such policies, price increases worldwide would have been smaller. This research paper also concluded that Brazil's sugar cane based ethanol has not raised sugar prices significantly, and recommends removing tariffs on ethanol imports by both the U.S. and E.U., to allow more efficient producers such as Brazil and other developing countries, including many African countries, to produce ethanol profitably for export to meet the mandates in the E.U. and U.S.

An economic assessment report also published in July 2008 by the OECD agrees with the World Bank report regarding the negative effects of subsidies and trade restrictions, but found that the impact of biofuels on food prices are much smaller. The OECD study is also critical of the limited reduction of GHG emissions achieved from biofuels produced in Europe and North America, concluding that the current biofuel support policies would reduce greenhouse gas emissions from transport fuel by no more than 0.8% by 2015, while Brazilian ethanol from sugar cane reduces greenhouse gas emissions by at least 80% compared to fossil fuels.

The assessment calls on governments for more open markets in biofuels and feedstocks in order to improve efficiency and lower costs.A study by the Brazilian research unit of the Fundação Getúlio Vargas regarding the effects of biofuels on grain prices.

Concluded that the major driver behind the 2007-2008 rise in food prices was speculative activity on futures markets under conditions of increased demand in a market with low grain stocks. The study also concluded that expansion of biofuel production was not a relevant factor and also that there is no correlation between Brazilian sugarcane cultivated area and average grain prices, as on the contrary, the spread of sugarcane was accompanied by rapid growth of grain crops in the country.

FOOD PRICE INFLATION: CAUSES AND IMPACTS

U.S. food prices are increasing. According to USDA, the Consumer Price Index (CPI) for "all food" increased 4% in 2007, the largest annual jump since 1990. In 2008, this trend is expected to continue: the "all food" CPI is forecast to increase 3.5% to 4.5%. This rapid inflation follows an extended period of stable food prices. From 1987 through 2007, food prices increased an average of 2.7% per year, excluding the drought years of 1989 and 1990.2 During 2005 and 2006, food prices rose 2.4%. This report examines the cause of food price increases and evaluates their impacts on U.S. consumers.

KEY FACTORS BEHIND HIGHER COMMODITY PRICES

Robust Domestic Demand. Corn, soybean, and wheat prices all reached 10-year highs during the 2006-2007 crop year. High prices for corn reflected increased use for ethanol (22% of the 2007 crop) and strong exports. High corn prices in turn encouraged growers to move acres from wheat and soybeans into corn, contributing to tight supplies and higher prices for those crops. U.S. farm prices in 2007 for corn are estimated at $3.75 to $4.00 per bushel, compared with $2.00 in 2005; soybean prices are estimated at $10.00 to $10.80 per bushel, up from $5.66 in 2005; and wheat prices are estimated at $6.50 to $6.80 per bushel, up from $3.42 in 2005.

Global Stocks Are at Low Levels. Globally, stocks of corn, wheat, and soybeans are at historically low levels. Drought in Australia and Eastern Europe and poor weather in Canada, Western Europe, and the Ukraine have reduced available quantities.

With world stocks for wheat at a 30-year low, 4 buyers are turning to the U.S. for supplies. Global Consumption Patterns Are Changing. Higher incomes are boosting demand for processed foods and meat in countries such as India and China. These shifts require more feed grains and edible oil. Even in low-income countries of sub-Saharan Africa, Asia, and Latin America, the vegetable oil share of diets has risen as processed food consumption rises. In China, consumption of meats, other livestock products, and fruits has increased while consumption of grain-based foods (such as bread) has slipped.

Improving food distribution systems are altering Chinese food preferences by introducing non-local foods. In India, per capita consumption of grains has fallen, while that of animal products, edible oils, vegetables and fruits has increased. Better food distribution systems are altering Chinese food preferences by introducing non-local foods.

Weak Dollar Boosts Demand for U.S. Exports. As the dollar depreciates against foreign currencies, U.S. exports become more competitive, boosting demand and prices. The dollar, adjusted for relative inflation rates, is expected to depreciate 7% against the euro, 6% against the Chinese yuan, and 8% against the Brazilian real in 2008.

The exchange rate is an important determinant of agricultural trade. The depreciation of the U.S. dollar since 2002 has helped improve U.S. Agricultural export performance. According to the U.S. Department of Agriculture's (USDA's) Economic Research Service (ERS), the dollar is forecast to be up 2% versus the yen, unchanged against the Canadian dollar, down 2% against the Mexican peso, and down 6% against the Argentine peso in 2008.

How Do Higher Commodity Prices Impact Consumers?

As commodity prices rise, food prices follow, but to a lesser extent. On average, about 20 cents of each dollar spent on food is the farm share — the retail cost less the value-added after the product leaves the farm and moves along the marketing chain to the retail outlet. In less processed foods, the farm component of the final product is larger and changes in the farm price have a greater impact at the retail level. For instance, eggs, and fresh fruits and vegetables undergo minimal processing after they leave the farm — they are consumed in essentially their original form. The retail value of such products tends to have a large farm component and changes at the farm level have a greater impact on the consumer.

On the other hand, in highly processed products, such as breakfast cereal, the corn, wheat, or rice used is completely transformed and the final product bears little resemblance to the original commodity. An 18-ounce box of corn flakes contained about 3.3 cents worth of corn in 2006.9 Higher corn prices in 2007 increased the corn share to 4.9 cents. This is a small part of the retail value of a box of corn flakes. Most of the retail price represents packaging, processing, advertising, transportation, profit, and other costs.

Energy Costs

Energy costs affect all levels of the food production sector. Recent record crude oil prices in excess of $110 per barrel affect costs throughout the marketing chain. Producers spend more for fertilizer (for which natural gas is a major input), crop drying, and transportation — raising production costs. At the processing, wholesale, and retail levels, the cost of transportation and operating packing houses, manufacturing plants, and retail stores has increased. Some of these costs are passed on to consumers in the form of higher prices. In addition, high petroleum prices increase the competitiveness of ethanol, further boosting demand for corn.

Food Price Changes by Food Type

Meat, Poultry, Dairy, and Eggs. The CPI for all meats advanced 3.3% during 2007.11 Beef increased 4.4%, pork 2%, broilers 5.2%, and eggs 29.2%, and dairy products advanced 7.4% in 2007. The farm share of these products is large compared with other foods, so changes at the farm level are passed, to a greater extent, to the consumer. In many cases, higher feed and energy costs were behind these increases. Strong export demand — spurred by the weak dollar — and reduced flocks played a role in the price hikes for poultry and eggs. The CPI for meats is forecast to increase by 1.5% to 2.5% in 2008. Compared with other food categories, these high-value items also account for a large share (11.1%) of the average consumer's food budget.

Fruits and Vegetables. Prices for fruits and vegetables gained 3.8% in 2007 and are forecast to increase 3% to 4% in 2008. Production shortfalls affected some varieties, especially bananas, the largest by volume. Supplies of oranges were strong, offsetting other declines.

Energy costs were a large factor in higher fruit and vegetable price increases. Fruits and vegetables account for 8.4 cents of the consumer food dollar.

Cereals and Bakery Products. The CPI for these items advanced 4.4% in 2007 and is projected to rise 6.5% to 7.5% next year. Tight global wheat supplies and acreage reductions to promote ethanol production have caused a spike in wheat prices. However shifts in wheat prices have a relatively small impact on grocery store prices because the farm share of these products is small. Prices for these product are affected more by marketing factors such as transportation, labour, and energy costs than the cost of basic inputs.

Oilseeds and Related Products. Low stocks and strong export demand for soybeans are reflected in the CPI for these products, which gained 2.9% in 2007. While much of this category is supplied by soybeans, substitutes exist and will help moderate increases. In 2008, the CPI is set to rise 7% to 8% due to continued strong export demand from countries where changing diets require more vegetable oil.

Impact on Low-Income Households

Although U.S. consumers generally spend a smaller share of their income on food compared with many other countries, that share varies widely across income levels.

Overall, U.S. households spend 12.6% of their income on food, so changes in the price of food have to be large to affect their total budget. However, the picture is vastly different for low-income households. In 2006, households with incomes in the lowest reported income category spent 17.1% of their income on food. Households with incomes greater than $70,000 spent 11.3% of their income on food.

When food prices rise, families with lower incomes feel the pinch more acutely since food expenditures make up a larger share of their total expenditures. Also, higher-income families can shift food consumption to the home from restaurants, saving money without reducing consumption. A 4% to 5% increase in food expenditures has a significant impact on purchasing power for low-income families.

Federal Spending for Domestic Food Assistance Programmes

Food price inflation increases spending on domestic assistance efforts. Increasing prices encourage those who are eligible but not participating to enroll. Increasing prices translate directly into benefit payments and per-meal subsidies for entitlement programmes in which benefits are indexed to food-price inflation (*e.g.*, food stamps, school meal programmes). Increasing prices place pressure on appropriators to provide more funding to support caseloads for discretionary programmes like the Special Supplemental Nutrition Programme for Women, Infants, and Children (the WIC programme). Food Stamps. The Food Stamp programme is the largest of the federally supported domestic food assistance programmes. Its benefits are indexed annually for changes in the cost of USDA's least costly food plan, the "Thrifty Food Plan" (TFP). For a number of years and well into 2006, annual increases in the TFP typically ranged between 1.5% and 2.5%, with a few exceptions. However, starting in late 2006, food prices (as reflected in the cost of items in the TFP) began to increase at a faster rate. The last benefit increase, effective October 2007, was 4.6%. As a result, the average monthly benefit will be $6 per person higher in FY2008.

The impact of benefit increases on food stamp costs also depends on participation. For FY2008, the benefit increase noted above (combined with estimated growth in enrolment) yields a likely $2 billion cost attributable to adjustments for food price increases (out of total spending of $36.7 billion), about double the $1 billion that would have occurred based on pre-2007 price increases. Costs are expected to increase even more in FY2009.

Child Nutrition. Federal payments for lunches and breakfasts served to children in participating school meal programmes are the second largest federal commitment to domestic food assistance, about $11 billion per year. These per-meal subsidies — now ranging as high as $2.83 a meal, including the value of USDA commodity donations — are indexed every July to food-price changes reflected in the "Food Away From Home" component of the CPI over the 12-month period ending each May.

Indexed maximum subsidy rates (those paid for the majority of school meals that are served free or at a reduced price to children from lower-income families) have increased by some 25 cents a meal between the 2005-2006 school year and the current 2007-2008 school year. The annual increase in subsidies has gone from 2.9% for the 2005-2006 school year to 3.3% for the 2007-2008 school year, increasing federal support by about $300 million above spending if earlier food price increases had prevailed. According to ERS, this trend is expected to continue into FY2009.

The WIC Programme. Unlike food stamps and child nutrition programmes, the WIC programme is discretionary. Spending depends on annual appropriations, based largely on estimates of participation and the cost of the food packages that are purchased with WIC vouchers. The value of benefits is not indexed, per se. Rather, WIC vouchers are redeemable at whatever the participating retailer charges for the items covered by the CRS-6 13 USDA/ ERS, Rising Food Prices Intensify Food Insecurity in Developing Countries," *Amber Waves*, February 2008. vouchers, which differ according to the type of recipient (*e.g.*, pregnant mother, infant, child). As a result, the cost of WIC vouchers reflect food price changes without the time lag built into other nutrition programmes like food stamps. Just as important, WIC vouchers are highly specific as to the food items they cover and have a relatively heavy emphasis on certain types of food — dairy items and infant formula being a major component.

In recent years, the cost of WIC food vouchers has varied a great deal, largely because of changes in dairy-related food prices. The average per-participant monthly cost of vouchers has ranged from $34.80 in FY2002 to $39.15 in FY2007. However the annual percentage increase has been very small for some years (1% or less for FY2003, FY2005, and FY2006) and more substantial for other years (6.6% for FY2004 and 5.6% for FY2007). Most recently, monthly per-participant WIC food costs averaged $42.50 for the first three months of FY2008. Given this significant volatility, it is difficult to produce specific estimates of the effect of food price inflation on WIC programme costs. However, the ERS forecasts of increases in egg and dairy product prices in the 2% to 4% range in 2008 indicate that relatively high WIC food costs are likely in the near term.

Foreign Food Aid

Higher commodity and food prices reduce our ability to provide food aid to other countries without additional appropriations. Food aid usually takes the form of basic food grains such as wheat, sorghum, and corn, and vegetable oil — commodities critical to developing-country diets. Since there is very little value added for these commodities, shifts in prices translate directly into higher prices for food-insecure countries or reduced food aid contributions per dollar spent.

Also, higher energy costs have increased shipping costs for both food purchases and food aid. Unlike some domestic nutrition programmes, foreign food aid is not adjusted to account for changing costs. After a long period of declining food costs, developing countries are facing increased food import bills — for some countries as high as 25% in 2007. The U.S. Agency for International Development (USAID) has indicated that rising food and fuel prices would result in a significant reduction in emergency food aid.

According to press reports in March 2008, USAID expects a $200 million shortfall in funding to meet emergency food aid needs. For FY2008, Congress appropriated $1.2 billion for P.L. 480 food aid, the same as FY2007. For FY2009, the President's budget again requested $1.2 billion. In six out of ten years since 1999, supplemental funding for P.L. 480 Title II food aid has been appropriated. Last year, the U.N. World Food Programme (WFP) estimated it would need $2.9 billion to cover 2008 food aid needs. Recent commodity, energy, and food cost increases have boosted this estimate to $3.4 billion. According to the WFP, the current price increases force the world's poorest people to spend a larger proportion of their income on food.

CAMPAIGNS TO CONTROL PRICES AND PRODUCTION

In the past few decades governments have undertaken to control both prices and output in the agricultural sector, largely in response to the pressures of the farmers themselves. In the absence of such control, farm prices tend to fluctuate more than do most other prices, and the incomes of farmers fluctuate to an even greater degree. Not only are incomes in agriculture unstable, but they also tend to be lower than incomes in other economic sectors.

THE PROBLEM

The instability of farm prices results from several factors. One is the relative slowness with which farmers are able to respond to changes in the demand for their product. Farmers generally must produce on the basis of expectations, and if their expectations turn out to be wrong, the resulting surplus or shortage cannot be corrected until the beginning of the next production cycle. Once a crop is planted, very little can be done to increase or decrease production in response to market prices.

As long as prices cover current operating costs, such as the cost of harvesting, it pays farmers to carry through their production plans even if prices fall to a very low level. It is not unusual for the prices of particular farm products to vary by a third or a half from year to year.

This extreme variability results from the relatively low responsiveness of demand to changes in price—*i.e.,* from the fact that in order to increase sales by 5 percent it may be necessary to reduce the price by 15 percent. The instability of farm prices is accompanied by instability of farm income.

While gross income from agriculture generally does not vary as much as do individual farm prices, net income may vary more than prices. In modern agriculture costs tend to be relatively stable; the farmer is unable to compensate for a drop in prices by reducing his payments for machinery, fertilizer, or labour.

The incomes of farm workers are generally below those of other workers. There are two major reasons for this inequity. One is that in most economies the need for farm labour is declining, and each year large numbers of farm people, especially young ones, must leave their homes to seek jobs elsewhere. The difference in returns to labour is required to bring about this transfer of workers out of farming; if the transfer did not occur, farm incomes would be even more depressed.

The second major reason for the income differences is that farm people generally have less education than do nonfarm people and are able to earn less at nonfarm jobs. The difference in education is of long standing and is found in all countries, developed and undeveloped; it also exists whether the national education system is highly decentralized, as in the United States, or highly centralized, as in France.

Government Intervention

Governments have employed various measures to maintain farm prices and incomes above what the market would otherwise have yielded. These have included tariffs or import levies, import quotas, export subsidies, direct payments to farmers, and limitations on production. Tariffs and import quotas can be effective only if a nation normally imports some of its supply. Export subsidies result in higher prices to domestic consumers than to foreign purchasers; their use requires control over imports to prevent foreign supplies from entering the domestic market and bringing prices down. Direct payments to farmers have been used to maintain prices to consumers at reasonable levels, while assuring farmers a return above world-market levels. Limitations on production, intended to reduce supply and thus increase prices, have been used mainly in Brazil (for coffee) and in the United States (for major crops).

Since the enactment of the Agricultural Adjustment Act of 1933, the United States has had programmes designed to limit the production of major farm crops through restrictions on acreage. Since that date it has also offered price supports for major crops such as wheat, feed grains, rice, tobacco, peanuts (groundnuts), and cotton, as well as for manufactured dairy products.

It has not had price-support programmes for perishable crops or for major livestock products except for a few years during and after World War II. The price-support method most widely used has been the nonrecourse loan made by the Commodity Credit Corporation (CCC): the farmer may repay the loan by delivering his produce at the support price or "redeem" it in cash if the market price is higher. The amount of particular crops offered for price-support loans has varied greatly from year to year, as have redemptions.

Most of the farm products given price supports were crops normally exported by the United States. Until the mid-1960s the price supports were above the export prices. Unless export subsidies were paid to make up the difference between domestic prices and the prices foreign buyers were willing to pay, exports became impossible.

Export subsidies were accordingly paid on such farm commodities as cotton and grains. In the 1960s the support prices for the major export commodities, except tobacco, were established at levels near or slightly below world prices to permit market forces to manage the distribution of supplies between domestic and foreign markets. The lowering of support prices was accompanied by a substantial increase in the size of direct payments to farmers. By the end of the 1960s such payments had come to constitute a high percentage of the cash receipts from farm marketings: in cotton, 60 percent; in wheat, 40 percent; and in feed grains, 30 percent. These payments fell sharply in the late 1970s, largely as a result of increased demand.

In order to receive payments, farmers had to agree to limit the acreage devoted to specified crops. At the beginning of the 1970s the various programmes had resulted in the diversion of approximately 20,000,000 hectares of land from the production of major farm crops. The number of acres diverted from cultivation fell sharply, however, in the late 1970s and early 1980s.

A major component of U.S. farm price policy since World War II has been the disposal of surplus produce abroad through the economic aid programme. This began as an outgrowth of wartime Lend-Lease, and food exported from the United States made a major contribution to the postwar recovery of western Europe. The Agricultural Trade and Growth Act of 1954 provided a base for continuing such activities, and gradually the emphasis shifted from western Europe to the developing countries. One of the important effects was to dispose of farm products that could not be sold either domestically or in regular commercial foreign trade. Without this the farm income and price objectives could not have been achieved except by more stringent output limitations, lower farm prices, and larger direct government payments.

Efforts to control agricultural prices go far back in English history, although the early objectives were quite different from those of more recent times. The Corn Laws of the 15th century were designed to prevent prices from becoming too high; restrictions were imposed on the right to export corn (wheat) when the domestic price exceeded a specified level. In 1663 the laws were revised to prevent prices from falling too low, by including import duties when the home price did not exceed a specified level.

The general trend, until the Corn Laws were finally abandoned in 1846, was increasingly towards ensuring higher prices for home producers through the payment of export bounties and by the restriction of imports until prices reached specified levels. After 1846 the British followed a free-trade policy for agricultural products but moved to the protection of agriculture and the establishment of minimum prices for certain farm products during the depression of the 1930s. Protection was expanded after World War II by legislation in 1947 and 1957 which sought to support farm prices primarily through deficiency payments to farmers, covering about 95 percent of total output. In most cases the domestic price was free to vary with changing demand and supply conditions; local products competed with imported supplies that were generally subject to relatively low tariffs. The farmer was reimbursed for the difference between his average realized price and a guaranteed price. The Agricultural Act of 1957, which gave the government the right to limit the amount of agricultural output on which deficiency payments were made, was designed to reduce the cost of the programme and to encourage domestic production.

The British system of supporting farm prices, while allowing consumers the lowest possible food prices in the world market, was gradually abandoned during the late 1960s as the United Kingdom prepared for entry into the European Economic Community (EEC).

When the United Kingdom entered the EEC in 1972, its agricultural prices began to rise to the much higher level prevailing within the EEC. The United Kingdom, moreover, imports more food and live animals from EEC countries than it exports, leading many British to question the value of membership in the EEC.

The EEC has established a common agricultural policy (CAP) for the Common Market countries. The CAP, worked out for each major farm commodity, was originally designed to create free trade for that commodity within the community. Special subsidies by the individual countries, and other national farm programmes, were to be eliminated to prevent competitive advantages. The first of the regulations implementing the CAP were enacted in 1962 and applied to grain (except rice), poultry, eggs, live hogs and whole hog carcasses, fruit and vegetables, and wine. Similar programmes were developed later for beef, dairy products, sugar, rice, and fats and oils.

The most important features of the CAP mechanism are the target prices, the threshold prices, the support or intervention prices, the variable levies on imports to make up the difference between landed prices and threshold prices, and export subsidies or refunds equal to the difference between market prices in the EEC and in the importing country.

For most CAP commodities the primary device for achieving target prices is the variable import levy. This levy, which fluctuates with the import cost of a commodity, keeps the domestic price at or near the target price if the commodity is imported. When EEC production of a commodity exceeds EEC consumption, the authorities may purchase the commodity for storage, pay to have it processed for another use (*e.g.,* wheat may be denatured and sold as a feed grain), or subsidize its export to countries outside the EEC. With these techniques the EEC has been able to maintain farm prices at levels substantially higher than those prevailing in the United States and Canada.

Throughout the 1960s the EEC did nothing to limit or control the production of agricultural products. When large stocks of butter and dry skim milk accumulated, and as the costs of maintaining dairy product prices and subsidizing wheat exports mounted, thought was given to reducing production. A payments programme to induce shifts from dairy to beef production was inaugurated, and there was talk of reducing the area cultivated for grain. Output limitation has been made difficult, however, by the significant differences in circumstances among the farmers in the various EEC countries.

SOVIET POLICIES

The farm policies of the Soviet Union were established during the First Five-Year Plan (1928–32), when agriculture was collectivized. For all practical purposes, regular markets for farm products were abolished at that time, and each collective farm was required to deliver an assigned quota of produce to the state at very low prices.

If a farm had anything left after meeting the obligatory quotas, it could sell the surplus to the state at higher prices or to the local free markets. Until 1958 the collective farms also had to make payments in kind to the machine tractor stations in return for work done. After Stalin's death in 1953, farm prices were increased significantly; the average procurement prices for food products increased almost fourfold between 1950 and 1956. In 1958 the multiple-price system was abandoned, and the prices paid to collective farmers became almost seven times the average paid in 1950. The machine tractor stations were abolished in 1958 and the machinery transferred to individual farms.

Another major revision of prices was made by the post-Khrushchev government in 1964–65. A two-price system replaced the single prices; prices for deliveries of grain up to the planned amount were about 10 percent higher than the previous single price, and deliveries above the plan level received a premium of 50 percent. Significant regional price differentials were established to cover the higher costs of production in some regions.

Prices of livestock products had already been increased by about 35 percent in 1962, and in 1965 further increases of perhaps a third were made. Another important measure was a commitment that planned purchases were to be fixed at specified levels during the Eighth Five-Year Plan (1966–70), both in the aggregate and for individual farms.

Prior to that time, if a farm had significantly increased its production or if other farms in the same region had failed to meet their deliveries, the delivery quota might be arbitrarily increased for the farm that happened to have had some output available for delivery.

Soviet price policy before 1953 was clearly designed to obtain farm products as cheaply as possible. The low prices were generally not passed on to consumers; a significant fraction of total governmental revenue was derived from high taxes on farm products.

The changes made after 1953 were intended to provide farmers with an incentive to raise production and to make more efficient use of resources. Only a part of the increase in prices paid to farms was passed on to consumers; much of the increase was at the expense of government revenue.

In the late 20th century Soviet planning began to give greater emphasis to private plots; while constituting only about 1.5 percent of Soviet farmland, these plots produced about one-third of the nation's agricultural output other than grains. Restrictions on the crops private plots could produce were relaxed, and the importance of those plots was stressed.

State farms and collectives, however, continued to receive the vast majority of capital and feed grains. Private plots, moreover, suffered from many of the problems that stunted the state farms and collectives, including the flight of young people from the countryside.

Accomplishments

The effects of price and income policies are difficult to assess. The policies have unquestionably worked to raise agricultural production in the countries where they have been applied, but their usefulness as a means of enhancing the economic well-being of farm people is debatable.

The governments of the industrial countries have been able to raise the returns from agriculture above the levels that would have prevailed in the absence of such intervention.

In addition to maintaining prices, they provide subsidies for agricultural inputs such as tractor fuel and chemical fertilizers; they also gave assistance in consolidating small farms into larger ones and in improving farm buildings. They do not, except for the United States, attempt to moderate the effects of these policies on production.

The level of income and the economic well-being of farm people in general are determined by many factors, including not only the prices they receive for their output but also the rate at which the economy in general is growing, the ease with which people can move from farm to nonfarm jobs, the prices they must pay for their productive inputs, and their level of education.

With respect to average income per person, as distinguished from total income, the prices received and paid are probably less important than the other factors mentioned.

This becomes obvious when one compares farm incomes in the United States or the United Kingdom with those in Argentina or India; the differences in real income have to do mainly with the levels of economic growth and not with farm prices or subsidies.

Government efforts to increase farm prices are likely to be offset, in the long run, by an increase in the number of persons engaged in farming, and this tends to keep the returns to farm labour from rising much faster than they would in the absence of such policies.

There are two other reasons for believing that the income effects of higher farm prices or subsidies are relatively insignificant in the long run compared with other factors affecting incomes of farm workers. One is that an increase in farm prices induces farmers to use more fertilizer, machinery, fuel and oil, and other items. If a significant part of any increase in gross income is used for such things, the absolute increase in net farm income is much smaller than the increase in gross farm income.

The second reason is that a given increase in government-supported farm prices generally occurs only once. After the increase in returns has been realized, the higher farm prices contribute nothing further to incomes. In contrast, general economic growth along with the continued reduction of the farm labour force has cumulative effects on the return to farm labour.

If the returns to farm labour were to grow at an average annual rate of about 3 percent, for example, farm prices would have to increase at least 3 percent annually (assuming other prices did not change) to have the same effect on returns to farm resources.

The Problem of Standards

None of the governments engaged in regulating farm prices and incomes has been able to apply a meaningful standard as to what a fair price or reasonable income is. The actual measures adopted, such as specific price supports or intervention prices, have been determined through the political process, with little reference to formal principles or standards.

In the United States the Agricultural Adjustment Act of 1933 stated that the goal should be to establish prices having the same purchasing power as those of the period 1910–14. By the end of World War II it had become clear that the "parity price" relationships of 1910–14 were no longer relevant to existing conditions.

The Agricultural Act of 1948 retained the 1910–14 average as a parity for all farm products but stated that the parity for individual products was to be the average of prices over the most recent 10-year period. Since the application of this formula would have resulted in a significant reduction in the parity prices for some politically important farm products, particularly cotton and wheat, legislation that was passed in 1949 declared that the parity price for an individual commodity was to be determined by either the old or the new formula, whichever was higher. Not until 1955 were the "modernized parity" prices put into effect on a gradual basis. The Food and Agricultural Act of 1977 introduced cost of production as a standard for determining farm price supports. This standard, however, is far from absolute, since the cost of production varies from one region to another; moreover, many costs of production, such as rent, are influenced by the value of the crops produced.

There has been a similar lack of objective or measurable standards in other countries. In Great Britain, for example, the Agriculture Act of 1947 declared its intention to be that of promoting and maintaining a stable and efficient agricultural industry capable of producing such part of the nation's food and other agricultural produce as in the national interest it is desirable to produce in the United Kingdom, and of producing it at minimum prices consistently with proper remuneration and living conditions for farmers and workers in agriculture and an adequate return on capital invested in the industry.

Several other countries have legislation that aims, without specifying in practice what is meant, to obtain for farm people the same level of income as that of other groups in the economy or that states that farm people should share in the rise in real per capita incomes. Finland, Japan, France, Sweden, and Norway have such policy objectives. German legislation declares that agriculture should share in the progressive growth of society and requires that the government each year prepare a report showing the extent to which the return to farm labour, or properly managed holdings under average conditions, is in line with that of wage earners in comparable non-agricultural occupations in rural areas.

The agricultural price objectives of the Treaty of Rome, which established the EEC, also lack practical significance: To ensure... a fair standard of living for the agricultural population, particularly by the increasing of the individual earnings of persons engaged in agriculture; to stabilize markets; to guarantee regular supplies; and to ensure reasonable prices in supplies to consumers.

Costs

The costs of the agricultural price and income policies of industrial countries are substantial; they include not only direct governmental outlays but also the increased costs to consumers in those countries, as well as the losses to developing countries of potential export markets. The high prices of farm products in the United States in the mid-1970s and the relaxation of interventionist policies by the EEC after 1974 substantially reduced the costs of farm programmes in these two regions. With the decline of farm prices that began in 1976, costs to taxpayers and consumers again approached the levels of the early 1970s.

PRICE INFLATION AND THE AGRIBUSINESS INDUSTRY

For purposes of illustration, consider the following:

Robert Ask has been a longtime, good customer of your retail farm supply store. In fact, Bob's father had served as a member of your store's first Board of Directors at the time of its organization some 27 years ago. As you glance out your office window this particular morning, you see Bob driving his pickup into the loading area on the south side of your store's warehouse.

It's a pattern which has been repeated every 8 to 10 weeks for as long as you can remember: *i.e.*, Bob has driven in to buy his usual load of cattle feed supplement, perhaps a salt block, and maybe some other small items like an oil filter for his truck or a gallon of antifreeze. On this particular morning, you are less than anxious to engage in the usual chitchat with Bob. The reason is painfully evident in the price you will be charging Bob for his supply of feed. Earlier this week you were forced to increase, dramatically, all your mixed feed supplement prices. Moreover, this was the third such increase in the past three months. Despite your fears and within moments, you turn around at your desk to find Bob's reddened face looking at you through your office door. The normal amenities of a greeting are cast aside as you begin to deal with the rapidfire sequence of questions directed at you by your maddened friend and longtime customer.

While the above episode has been totally fabricated, its real life counterpart has been reenacted time and time again throughout the agribusiness industry this past year. No doubt you too have been subjected to numerous questions from anxious customers. In some cases you may even have been the recipient of some customer heckling or direct emotional outrage. Never before has the agribusiness manager been more severely tested by the reactions of his clients and/or customers to the direct effects of inflation. Chances are that for many residents of the rural areas, your business and the local grocery store represent their total perception of how inflation affects them. As a result, your confused and disenchanted customers look to you for an answer to such concerns as:

"What has Caused the Rapid Price Increase?"

"Why must I be forced to carry the burden of higher costs while large corporate profits grow to record levels?" "How can inflation be combated?" and "What can I do to minimize the detrimental impact of inflation?" Perhaps you have already developed a stock answer to customer's questions such as these. If not, the purpose of this letter is to provide answers, where they exist, to such plaguing questions.

Obviously, simple answers do not always exist. At times, more than one answer can be secured from a single question. In short, the phenomenon so glibly referred to as "inflation" is a very complex process. Its causes, effects, and cures are not thoroughly understood by even the brightest economists. As a result, this author can hardly come forth with any startling or innovative declarations. Instead, an attempt shall be made to provide the reader with a better understanding of that which is both known and agreed to about inflation.

Inflation—What is It?

In an overly simplified sense, economists might choose to define inflation as "too much money chasing too few goods." The standard textbook definition might be "inflation is an increase in the volume of money and credit relative to available goods, resulting in substantial and continuing rise in the general price level."

So if your customer asks, tell him that inflation, as evidenced by rapidly rising prices, is the result of the money-aftergoods chase. The purchasing power of the money surplus decreases and, as an end product, the Robert Asks of this country pay higher prices for cattle feed. The traditional monitors over the inflationary rate are the Consumer Price Index (CPI) and the Wholesale Price Index (WPI).

Each is a measure of the general price level and is reported monthly. While the WPI concentrates on wholesale prices of resources, services, and products, the CPl concentrates on prices at the retail level.

Each has shown a continuous rise over recent years and a much more dramatic increase in recent months. At the present time (1974), the CPI is rising at an annual rate of 12 to 14 percent, while the WPI is exceeding 20 percent on an annual basis. Perhaps worst of all, there are few signs that this rate will diminish significantly in the next few months.

Inflation-What are its Causes?

It should now be apparent that inflation is not the result of a single flaw in our economy. The causes of inflation are rooted deeply to three fundamental patterns: 1) the set of tastes and preferences we have become accustomed to as a consuming society, 2) the structure of the economy as we have shaped it, and 3) the economic policies that we have adopted in our attempt to mold both 1 and 2 above. Let's now look at each of these fundamental patterns. Many of the ideas presented here are taken from "Inflation: Some Sources and Effects." Virginia Agricultural Economics, by J.P. Marshall, September, 1974.

Our tastes and preferences. Over the years, we have always been faced with thc choice between the risk of recession and the risk of inflation. We have always opted in favour of the risk of inflation. The reasons are quite clear. Recessions are politically unpleasant.

A recession means unemployment and a general drop in the total economy's level of output. In brief, our total output of goods and services grows more slowly than does our level of tastes and preferences for these goods and services. To attain a low level of unemployment and a moderate annual increase in GNP, we have chosen to encourage a high rate of borrowing and a low rate of saving. The truth is; our consuming psychology is such that we are quite willing to accept annual price increases as long as an adequate supply of goods and services is maintained.

Our economic structure. That prices and wages have shown little tendency to adjust downward in recent periods is evidence of a major component of rigidity within our economy. Wages, in particular, are not permitted to decline due to the bargaining strengths of our unions. Management, however, is also a contributor because it operates under the assumption that wage increases can be passed directly to the consumer. In its initial stages, inflation thrives on itself. For example, an awareness of inflation tends to accelerate the rate of price increases as consumers stockpile their purchases in expectation of still further price rises. Labour immediately inserts escalator clauses into its contracts and lenders immediately insert variable interest rate clauses into their mortgages and other financial instruments.

Shortages caused by other structural problems within our economy also contribute to inflationary pressures. Agricultural products and oil are classic examples of shortages arising from rigidities in the economic structure. Other shortages, such as electrical energy, are the result of inadequate investment and capacity constraints which make it impossible to meet demand at the existing price.

Our economic policies. Economic policy is not created in an arbitrary manner. In fact, it is created in direct response to the existing consumer preferences and our economic structure. Efforts to control inflation are most often focused on a reduction of government expenditures or a decline in the rate of growth in the money supply. While these efforts may have a moderating influence on inflation, adjustments made only in these two areas will not control inflation. To do so would require some-perhaps major-changes in our system of consumer preferences and the adoption of policies that will alter the economic structure and, thereby, dampen inflationary tendencies on a more permanent basis.

Inflation—Two Major Types

Classical economic thought proposes that there are two major types of inflation, "Demand-Pull Inflation" occurs when there is an increase in the amount of money available to buy a fixed supply of goods.

This situation was very much in evidence when U.S. food prices raced dramatically upward in 1973 after nearly two decades of relative stability. "Cost-Push Inflation," on the other hand, is largely the result of rises in costs (especially labour) not offset by increases in productivity. As a result of this situation, the general price level is forced upward.

Let's now consider some of the particulars of our agricultural economy for the purpose of identifying which of the two types of inflation now exists. Background considerations. To provide some perspective to the origin of price inflation, as it now exists, we shall consider some historical facts as they relate to U.S. agriculture. First of all, the price elasticity of export demand for feed grains is such that a 1.5 percent increase in quantity exported will result in a one percent decrease in price received. Second, we find that the markets for agricultural products are growing at varying rates, with vegetable oils and beef experiencing the most rapid market growth. Third, livestock is the principal means by which a large share of our crop production may be utilized. Given these three items, during the late 1960's and early 1970's our feed grain production grew by about 25 percent, while our exports of feed grains grew by nearly 50 percent. Markets for other U.S. farm products also expanded, and domestic food prices were kept stable via the market regulating effects of substantial grain carryovers.

In 1972, however, the Russians suffered a severe drought. The Russian government, therefore, was confronted with the choice of reducing the number of grain-consuming livestock or initiating a massive grainbuying programme. As we all know, it was the latter action, which ultimately led towards our involvement in the now infamous "Wheat Deal."

Following the sale of grain to Russia, our carryover stocks were all but eliminated and along with it the loss of the price stabilizing influence. The subsequent rise in feed grain prices to their current record levels is now well known by all.

On the demand side, perhaps the most important influence causing inflation in food prices was the change in consumer diets in both the United States and other countries. Quite simply, people in Western Europe, Japan, and here at home shifted from consuming grain directly to consuming it indirectly in the form of red meats and poultry. Thus, people's preferences for improved diets were consistent with their rising incomes and this resulted in increased pressures on the world stocks of grain.

Even as diets change, a basic economic fact must be acknowledged, *i.e.*, the price elasticity of demand for food is low. In other words, when food shortages occur, relatively large price increases may be expected. We long suffered under the illusion that our productive capacity was nearly unconstrained. Such was painfully untrue as evidenced by our most recent production records. World food shortages became known and, domestically, our food prices rose four times as rapidly during 1973 as prices of nonfood items. As a result, our policy makers are now confronted with a type of inflation that traditional monetary and fiscal tools are unable to deal with.

Accentuated by our own production restraints due to unfavourable weather conditions, world stocks of grain have now reached a level so low that a significant shock to the total production system could no longer be absorbed by storage supplies.

So, who is to be Blamed?

At a recent Food and Agriculture Conference on Inflation, each of the 40 conference participants (mostly agricultural economists) were surveyed as to their views regarding the major causes of 2 Source: Jerome M. Statter and Loren M. Solnik, "Effects of Escalators on Wages in Major Contracts Expiring in 1974," Monthly Labour Review, Washington D.C., U.S. Bureau of Labour Statistics, July 1974, p. 31. inflation. The results of this survey are of value for two reasons. First, they do confirm that some consensus is forming amongst professionals as to what the major contributing causes are. Second, they serve to illustrate that the list of causes is both long and varied. For example, deficit spending by government was the cause most often blamed and rated as a major cause by 70 percent of the respondents.

Supply shortfalls were also considered to be a significant cause. Natural causes, such as droughts and floods, were rated as a major cause by 45 percent of the respondents and shortages caused by economic and political action, for example, the oil embargo, were rated as a major cause by 63 percent.Concentration of power in unions and businesses, inefficient government programmes, overly ambitious monetary policy, and unfavourable weather were other causes listed, but by a much smaller percent of the respondents.

Cures to the inflationary process are not unknown. The taproot of inflation can be cut either by increasing the quantity of goods available or by reducing the volume of money spent and credit extended or by doing both. Unfortunately, the cures, once administered, are often painful and may cause side effects nearly as damaging as the illness itself.

It seems clear that efforts to remove inflation in food prices by reducing the domestic volume of money and credit extended are of no value. Only the slow process of rebuilding world grain stocks will moderate this source of inflation. However, as such stocks are rebuilt to abundant levels, U.S. food prices will go down and our agricultural economy may be unable to adjust.

Labour productivity continues to decline. As a result, income is redistributed, and the economic position of workers lacking bargaining power worsens as they appear among the growing ranks of the unemployed. For every one-tenth of one percent increase in the nation's unemployment rate, some 90,700 people have lost a paying job.

Current studies show that stable prices will be achieved when the unemployment rate reaches 8 percent (current level is around 6.7 percent). If you are unmarried, between the ages of 16 and 19, and black, the odds are only about 50:50 that you will be able to secure and retain employment during the adjustment period. The major unanswered question is whether or not the people of this country are willing to accept the burdens and inequities resulting from the total corrective process.

Bibliography

A. Rashid: *Introduction to Genetic Engineering of Crop Plants : Aims and Achievements*, I.K. International Publication, Delhi, 2009.

Anil Kumar Verma: *Labour Migration In Agriculture Sector*, Discovery Publishing House, Delhi, 2011.

B Politycka and C L Goswami: *Plant Physiology : Research Methods*, Scientific Publication, Delhi, 2007.

Bidhan Roy and Asit Kumar Basu: *Abiotic Stress Tolerance in Crop Plants : Breeding and Biotechnology*, New India Publication, Delhi, 2009.

Edward G Mahin and Ralph H Carr: *Quantitative Analysis for Agriculture Science*, Srishti Publication, Delhi, 2006.

Edwin Cyrus Miller: *Plant Physiology : With Reference to the Green Plant (3 Vols-Set)*, Biotech Books, Delhi, 2005.

Edwin Oxlade: *Plant Physiology*, Viva Books, Delhi, 2010.

G. Kalloo, Mathura Rai, Major Singh and Sanjeet Kumar: *Heterosis in Crop Plants*, Researchco Book Center, Delhi, 2006.

Herbert Kendall Hayes and Ralph John Garber: *Breeding Crop Plants*, Asiatic Publication, Delhi, 2008.

Iqbal Hussain: *Plant Physiology*, Oxford Book Company, Delhi, 2008.

Kabita Dutta: *Plant Physiology*, International Scientific Publication, Delhi, 2011.

M. Prakash and K. Balakrishnan: *Abiotic Stress Tolerance in Crop Plants*, Satish Serial Publishing House, Delhi, 2014.

Madan Pal Singh, Sangeeta Khetarpal, Renu Pandey and Pramod Kumar: *Climate Change : Impacts and Adaptations in Crop Plants*, Today and Tomorrow Publication, Delhi, 2011.

Mohr: *Plant Physiology*, Springer, New York, 2003.

N Bhupendro Singh: *Capital Formation in Agriculture Sector of Manipur (1972-1997)*, Akansha Publication, Delhi, 2007.

Neeru Mathur: *Plant Physiology*, RBSA Publishers, Jaipur, 2010.

P C Trivedi: *Plant Physiology : Current Trends*, Pointer Publication, Delhi, 2007.

P.B. Bansal: *Bioanalysis and Biosensors in Agriculture Science*, Gene Tech Books, Delhi, 2006.

Raj Patel: *Stuffed and Starved: What Lies Behind the Hidden Battle for the World's Food System*, HarperCollins, 2003.

Ranjay Kumar Singh, Nancy J.Turner, Victoria Reyes-Garcia and Jules Pretty: *Social Ecological Diversity and Traditional Food Systems: Opportunities from the Biocultural World*, NIPA, 2014.

Renuka Desai: *Plant Physiology*, Adhyayan Publication, Delhi, 2008.

Sandhya Agrawal: *Bioanalysis and Biosensors in Agriculture Science*, Oxford Book Company, Delhi, 2009.

Susheela M. Das and Abha Bhardwaj: *Plant Physiology*, Wisdom Press, Delhi, 2012.

U.D. Chavan and J.V. Patil: *Objective Agriculture Science*, Daya Publication, Delhi, 2011.

V.L. Chopra and R.S. Paroda: *Approaches for Incorporating Drought and Salinity Resistance in Crop Plants*, New India Publishing Agency, Delhi, 2015.

Venkatesh Bhardwaj: *Emission of Green House Gases (GHG) from Agriculture Sector and Measures for their Mitigation*, Agrotech Publishing Academy, Delhi, 2009.

Index

H

I

L

M

N

O

P

R

S

T

V

W